Case Study Research

The Quick Guide Series

Dan Remenyi

Case Study Research

First published: April 2012
Second Print: February 2013
Second Edition: July 2013

ISBN: 978-1-909507-17-3
Copyright © 2012 The author

All rights reserved. Except for the quotation of short passages for the purposes of critical review, no part of this publication may be reproduced in any material form (including photocopying or storing in any medium by electronic means and whether or not transiently or incidentally to some other use of this publication) without the written permission of the copyright holder except in accordance with the provisions of the Copyright Designs and Patents Act 1988, or under the terms of a licence issued by the Copyright Licensing Agency Ltd, Saffron House, 6-10 Kirby Street, London EC1N 8TS. Applications for the copyright holder's written permission to reproduce any part of this publication should be addressed to the publishers.

Disclaimer: While every effort has been made by the editor, authors and the publishers to ensure that all the material in this book is accurate and correct at the time of going to press, any error made by readers as a result of any of the material, formulae or other information in this book is the sole responsibility of the reader. Readers should be aware that the URLs quoted in the book may change or be damaged by malware between the time of publishing and accessing by readers.

Published by: Academic Conferences and Publishing International Limited, Reading, RG4 9SJ, United Kingdom, info@academic-publishing.org

Printed by Berforts Group
Available from www.academic-bookshop.com

Acknowledgement

My understanding of case study research has developed steadily over the past 20 years when I first used this technique myself. I have learnt much from many research degree candidates whom I have supervised, mentored or examined. One of the great outcomes of being involved in academic research is that we can learn from each other's experiences.

However I would like to point out my debt to Dr Paul Griffiths who has made a special contribution to my thinking in this book. Paul is a highly skilled practitioner in the art of Case Study Research and he has made a number of important suggestions which have been incorporated in this book.

Case Study Research

Contents

Acknowledgement i
Contents i
Preface to the Second Edition vii
Preface to the First Edition viii
Chapter 1 - What constitutes Case Study Research? 1
    1.1.   Observation as a basis of research 1
    1.2.   Definitions of a case study 2
    1.3.   Research methodology and research methods 5
    1.4.   Qualitative research 5
    1.5.   Hypotheses and theory development 6
    1.6.   A fuller definition of case study 6
    1.7.   A story is central to a case study 7
    1.8.   The scientific logic of the case study 9
    1.9.   Summary and conclusion 14
Chapter 2 - Different Types of Case Studies and Choosing Appropriate Situations to Study 15
    2.1.   Different meanings of the term case study 15
    2.2.   Teaching – learning case study 15
    2.3.   The research case study 19
    2.4.   Descriptive case studies or exploratory case studies .. 20
    2.5.   A one case study research project 22
    2.6.   More than one case study research project 23
    2.7.   The unit of study – the organisation is the unit of study 24
    2.8.   Multiple units of analysis 26
    2.9.   Suitability and gatekeepers 28
    2.10.   Suitability Profile for a Case Study 29
    2.11.   Summary and conclusion 32

Chapter 3 - Data: Quantitative and Qualitative ... 33
- 3.1. The importance of data ... 33
- 3.2. A definition of data ... 33
- 3.3. Do, dare, dedi, datum ... 36
- 3.4. Unit of analysis ... 37
- 3.5. Quantitative and qualitative data ... 38
- 3.6. Different realities or different lens ... 41
- 3.7. Natural and elicited data ... 43
- 3.8. Hard and soft data ... 44
- 3.9. Data can be overwhelming – data overload ... 45
- 3.10. Commonly available sources of qualitative data ... 48
- 3.11. Combining quantitative and qualitative data ... 52
- 3.12. Data Acquisition Plan ... 53
- 3.13. Summary and conclusions ... 54

Chapter 4 - The research proposal and protocol ... 57
- 4.1. Understanding and planning academic research ... 57
- 4.2. The need for flexibility ... 58
- 4.3. Organisation and planning ... 59
- 4.4. The Research Proposal ... 60
- 4.5. The research protocol ... 69
- 4.6. The parts of a research protocol ... 70
- 4.7. Part 1 - Finding the starting point ... 71
- 4.8. Part 2 – Establishing and collecting the data ... 72
- 4.9. Part 3 - After data collection is complete ... 75
- 4.10. Part 4 - What does the data mean ... 79
- 4.11. Part 5 - What the research delivers to the community ... 81
- 4.12. Part 6 - The research ends ... 82
- 4.13. Summary and conclusion ... 83

Chapter 5 - Collecting the data ... 85
- 5.1. Data collection - a major challenge ... 85
- 5.2. Empirical research ... 86
- 5.3. Opportunistic attitude to data collection ... 87
- 5.4. Log all data collection activities ... 87

Contents

5.5.	Data collection is not easy	89
5.6.	First contact with the organisation	90
5.7.	Mixed data	91
5.8.	Data access	92
5.9.	Sources of data	94
5.10.	Triangulation	97
5.11.	Reflections	98
5.12.	A pilot case study also referred to as a field test	100
5.13.	Cross case analysis	101
5.14.	Summary and conclusion	103

Chapter 6 - Data Analysis for a Case Study 105

6.1.	The art of the possible	105
6.2.	The jigsaw puzzle metaphor	105
6.3.	Answer the research question	106
6.4.	The tool box for analysis of case study data	108
6.5.	Creating a convincing case study narrative	108
6.6.	Producing lists of constructs and concepts	109
6.7.	Demonstrating the relationships	111
6.8.	Hypothesis testing	112
6.9.	Theoretical conjecture development	112
6.10.	Data analysis begins	113
6.11.	One case at a time	114
6.12.	Finding suitable case studies	115
6.13.	Writing up the case study	117
6.14.	Data overload	117
6.15.	Using one holistic case study research for theory creation	118
6.16.	One holistic case study research for hypotheses testing	125
6.17.	Multiple case study research for theory creation	126
6.18.	Summary and conclusions	132

Chapter 7 - The case study writer as a story teller - some style and form issues 135

7.1.	Introduction	135

7.2.	The importance of the write up	135
7.3.	The transcripts are not the case study	136
7.4.	Writing as a craft skill	136
7.5.	The narrative nature of the case study	137
7.6.	Styles of the narrative	138
7.7.	Important dimension of a case study write up	139
7.8.	Written in an engaging manner	140
7.9.	Care with language	141
7.10.	Different story frameworks or forms	142
7.11.	A chronology or timeline approach	143
7.12.	A play format	144
7.13.	A biography	144
7.14.	A recollection of voices	145
7.15.	Not mutually exclusive	145
7.16.	Summary and conclusion	146

Chapter 8 - The case study dissertation 148

8.1.	Case studies for research degrees	148
8.2.	Preparing the ground for the case study write up	148
8.3.	Chapter One – The Introduction	150
8.4.	Chapter Two – The Literature Review	150
8.5.	Chapter Three – The Research Design	151
8.6.	Chapter Four – Executing the Research Design	151
8.7.	Chapter Five – Conclusions of the Research	154
8.8.	Chapter Six – Reflections, Limitations and Suggestion for Future Research	155
8.9.	Appendices to the dissertation	155
8.10.	Summary and Conclusions	156

Chapter 9 - Pilot Studies or Field Tests for Case Study Research 157

9.1.	Introduction	157
9.2.	Refining the research design	158
9.3.	Data requirements	158
9.4.	Data collection	159
9.5.	Preparing data for analysis and interpretation	160

Contents

	9.6.	Checklist 160
	9.7.	Feedback 161
	9.8.	Summary and conclusion 161

Chapter 10 - Data Management for Case Studies – Make writing up easier... 163

- 10.1. Introduction 163
- 10.2. Filing systems for data saving and retrieval.............. 164
- 10.3. Some details with which the researcher is probably familiar 165
- 10.4. Directory and file names 166
- 10.5. Directories and subdirectories............................... 166
- 10.6. Merging files or combining data 168
- 10.7. Backing up data 168
- 10.8. Reference management software 171
- 10.9. Computer Housekeeping 171
- 10.10. Summary and conclusion 172

Chapter 11 - Ethics Approval for Case Study Research.......... 173

- 11.1. Introduction 173
- 11.2. Background to the importance of ethics approval 173
- 11.3. All research involving human participation 174
- 11.4. Procedural ethics 175
- 11.5. Ethics in practice 176
- 11.6. The proposed two outcomes of ethics approval 177
- 11.7. Documents need completion................................. 177
- 11.8. Corporate research approval 177
- 11.9. Ethics approval for case study research 178
- 11.10. Research Participant's Information Document.......... 179
- 11.11. A Letter of Informed Consent 179
- 11.12. Measuring instruments... 180
- 11.13. Getting going... 181
- 11.14. Summary and conclusion 181

Chapter 12 - Evaluating your case study research................. 187

- 12.1. Ways of evaluating a case study 187
- 12.2. Case study appropriate form and structure?............ 187

v

12.3.	Evaluating the form and structure of a case study	188
12.4.	Has the case study facilitated the answering of the research question?	191
12.5.	Summary and conclusion	193

Reference list 195

Useful URLs 198

Appendix A - A Short Note on Hypothesis Testing with Case Studies 199

Index 203

Preface to the Second Edition

Interest in case study research continues to grow as it becomes more obvious that a holistic approach to research questions can provide insightful results. However case study research is not always understood and sometimes researchers produce unstructured narratives which are not able to add something of value to the body of theoretical knowledge. Worse still some researchers who have elected to use an eclectic mixture of research methods feel that they can call their approach case study research. It is increasingly important to understand the scientific logic of case study research and to approach this type of research method systematically.

This book is in the Quick Guide Series and therefore does not claim to be a comprehensive treaty on case study research. Nonetheless it was decided in the Second Edition to add some new chapters which provide extra comment on the important issues which will concern anyone contemplating case study research.

Dan Remenyi PhD

July 2013

Preface to the First Edition

Using case study research for an academic degree or for research to be published in a peer reviewed journal is challenging and thus requires careful study and attention to much detail. It is not the intention of this book to provide that detail but rather to orientate readers toward what is required for this type of research. Further study will be necessary to acquire the necessary skill to become an accomplished case study researcher.

Although case study based research has been increasing in popularity over the past 20 years there is still a considerable amount of misunderstanding concerning what it can actually do and how it should be used in business and management studies.

There are a number of misconceptions that regularly arise and which cause confusion. One of the most commonly encountered is the suggestion that if a research project uses multiple sources of data or evidence it is then by default a case study. The second is that case study research is only valid in a qualitative environment. The third is that case study research is only of value at the early stages of research when the researcher is looking for an interesting research question. Those who espouse these opinions regarding the proper place of case study research generally refer to case studies as descriptive or exploratory.

There is also confusion about how many case studies are required for a creditable research project. I am reminded of my

Preface to the First Edition

own experience in 1988 when I started to work with case studies and I was told by a senior academic that for a doctorate I would need to complete 50 case studies. Few academics would hold this view today although more than a few would be uncomfortable about one case study or being able to comment on generalisability from case study research.

This Quick Guide does not attempt to be a definitive account of all the matters related to case study research. Such a task would require a large tome. This book addresses issues related to the definition of case study research; concepts related to the nature of data; the mixed data issues; the importance of clarifying what is to be done through the research proposal and protocol; the selection of case study locations; some concepts pertaining to case study analysis and writing up. Finally the book addresses how to evaluate a case study research project.

The book will help those who are considering using a case study approach decide whether this is for them and it will provide those who have already decided to take this route with a framework for understanding what is really involved. Case study research is too important in the field of business and management studies to be left to the haphazard approaches that are now sometimes used.

Dan Remenyi PhD
May 2012

Chapter 1
What constitutes Case Study Research?

1.1. Observation as a basis of research

From an empiricist's point of view new knowledge is discovered by exploration by which we mean examining artefacts or situations or events. This is primarily achieved through observation, in the broadest sense of the word, and by reflection on what it is we are observing and by discussing this with knowledgeable informants and colleagues.

Observation may be performed by collecting a variety of different types of evidence or data including numbers, text, images and other sensed stimuli (Gillham 2000). Those who focus primarily on numeric data refer to the research as being quantitative and those who focus on other types of data i.e. text, images and other sense stimuli refer to their research as being qualitative.

Quantitative and qualitative research are not as entirely different as some researchers believe. The differences are often overstated and the similarities regularly ignored. There can be material overlaps between these approaches to research (Punch 2005). Text and images can be used to support the processes of quantitative research, and numbers used to support qualitative research. Some researchers even argue that we should not talk about quantitative and qualitative research but should only say that there are quantitative and qualitative data. The more challenging the research question the more likely there will be a need to draw on both quantitative and qualitative data and techniques and thus it is good practice for researchers to make themselves comfortable with both approaches.

1.2. Definitions of a case study

Case study research allows challenging research questions to be addressed using multiple sources of data or evidence. The definition of a case study supplied by Yin (1989) is still probably the most useful and this is:

> *a case study may be defined as an empirical enquiry that investigates a contemporary phenomenon within its real life context, when the boundaries between phenomenon and context are not clearly evident, and in which multiple sources of evidence are used.*

There are five issues addressed in this definition and these are:

1. empirical enquiry;
2. contemporary phenomenon;
3. real life context;
4. boundaries are not clearly evident;
5. multiple sources of evidence.

It is worth noting that there is no comment made about whether or not the case study should have a quantitative or a qualitative research orientation as it is assumed that any

What constitutes Case Study Research?

relevant data may be used. These five characteristics may be interpreted as follows:

Empirical enquiry points out that a case study should be based on primary or sense based data and not be produced by only library or web based work involving secondary sources.

Contemporary phenomenon suggests that it is not appropriate to label research addressing historic events as being a case study. The question then is when should we consider the matter being researched to be 'historic'? As a possible heuristic a case study should address issues which are no older than, say, five years. However a case study could address background and contextual issues which extend much further back in time as long as the relevant data being studied is contemporary[1].

Real life context refers to the fact that a case study should not be based on an experimental setting where the environment in which the activities described are controlled. Real life means that we are studying situations over which the researcher does not have control.

Boundaries are not clearly evident points out that when studying complex or challenging research questions there are often a considerable number of variables at play and that it is not always obvious which variables are actually present and which variables should demand the researcher's attention. Some researchers refer to situations such as this as being 'messy' which simply means that the research does not

---

[1] It is possible that research which studies events dated further back in time could be labelled as historiography. The point being made here is largely a taxonomy issue.

have the clear cut focus that laboratory research normally does.

Multiple sources of evidence suggests that any data or evidence which can facilitate the understanding of and help us to answer the research question may be accessed and where appropriate employed. This statement has a very broad reach and needs some clarification as to whether there are any preferred data sources and this will be addressed later.

> **Difficulty of case study research**
>
> Some misinformed academic researchers will suggest that case study research is an easy option. This is not true and is only said by those who have not conducted quality case study research themselves.

Before examining the detail of the case study further it is worth noting that there is an even broader view as to the nature of the research processes incorporated under the name of a case study. Bell (1993) suggests that:

the case study approach is an umbrella term for a family of research methods having in common the decision to focus on an enquiry around a specific instance or event.

If we agree with Bell then case studies not only employ multiple sources of evidence but also multiple research methods. A first reading of this definition suggests that case studies will have a proclivity for multiple methods and if this is true then the definition of methods needs to be considered.

What constitutes Case Study Research?

1.3. Research methodology and research methods

The term research method is sometimes used interchangeably with research methodology which can lead to some confusion. Better practice is to use the term research methodology to refer to high level options available to researchers such as: whether to conduct the research using a quantitative approach and thus relying on positivist type assumptions or a qualitative approach and thus relying on interpretivist assumptions. The term research method is then used to refer to which data collection and data analysis techniques will be employed such as interviews, questionnaires, production statistics, focus groups, field studies or participant observer work just to name a few (Hamel et al 1993).

1.4. Qualitative research

It is important to note that there is no prohibition or direction in the definition of case study to prevent a quantitative approach being used. However by practice the term case study is not used when a quantitative piece of work is undertaken. So much is made of the differences between quantitative and qualitative research and individuals who support quantitative or qualitative approaches often engage in inappropriate acrimonious debate. The reality is that virtually all academic research projects in the social sciences require both quantitative and qualitative data. A simple example of this is that even when a questionnaire is to be used to acquire the data, the form and the content of the questionnaire can only be finalised after some qualitative discussions have been had in terms of the data requirements with which to answer the research question. Furthermore when the analysis of the data obtained from the questionnaire is complete, the process of understanding and interpreting the findings is essentially a qualitative research activity. Thus in a sense it may be said that there is some element of mixed data and even mixed method thinking in most if not all academic research projects in the social sciences. It seems that the combining of qualitative and quantitative data is necessary to produce synergies required for understanding and answering the research question.

1.5. Hypotheses and theory development

Another important issue is that case studies are versatile in the way they can be used and thus a researcher may use a case study to help test hypotheses and to develop a theory. Hypothesis testing, which is sometimes described as the hypothetico-deductive approach to research, uses classical hypothesis testing. This usually requires numeric data and the application of some statistical technique. In the case study environment hypotheses could be tested using small samples provided the claims for generalisability were carefully circumscribed.

Case studies can be used inductively to facilitate the creation of theory. Here data is collected and from close inspection and careful deliberation and reflection of the researcher a theory is postulated. This in turn may lead to hypotheses being articulated and tested. This illustrates that deduction and induction can be and often are coupled.

1.6. A fuller definition of case study

Returning to the definition of a case study and keeping in mind all these other issues a more complete version is:

A case study in academic research is a term used to describe a research initiative which is/has:

What constitutes Case Study Research?

1. Used to answer complex or challenging research questions;
2. An empirical approach to answering the research question;
3. Involving many variables not all of which may be obvious;
4. Qualitative, quantitative or mixed methods and can be used in either the positivist or interpretivist mode;
5. Presented as a narrative as a way of facilitating the answering of the question;
6. A clear-cut focus on a unit of analysis;
7. Recognised the context in which the research question is put and the answer is sought;
8. Not extended for a long period of time i.e. does not compete with historiography;
9. Enriched by multiple sources of data or evidence in order to offer a degree of triangulation.

Each of these nine dimensions is important and the research needs to satisfy in some part all of the nine dimensions. No one dimension is overwhelmingly more important than the others. However the fifth point which is "Presented as a narrative" is often not well understood and thus poorly addressed by researchers.

1.7. A story is central to a case study

A case study needs to have a specific story line which flows smoothly from the introduction of the research project, through the background of the topic as demonstrated in the literature review, to the methodology, to the data collection and so on. The story has to be unfolded carefully and logically as it is the story which will eventually constitute the argument that the researcher has added something of value to the body of theoretical knowledge. The main characteristic of the story is that it needs to be accurate and complete which means that all the aspects of an academic research project are faithfully described. The story needs to describe the logical path followed by the researcher as he or she finds his or her way through the maze of the research. The story needs to account for the academic decisions made and the basis on which they were made. One of the more important aspects of the narrative is the description of the methodology chosen and why it was chosen. Sometimes the question is asked, "How can story telling become the basis of a piece of scientific research?" The answer to this is both short and simple and it has not been expressed better than the words used by Gould (1997):-

Humans are story telling creatures pre-eminently. We organise the world as a set of tales.

Perhaps Gould might have added that this penchant for story telling is no less in research and science than in any other aspect of life. Stories told well deliver understanding and knowledge in every aspect of life. Shattuck (1996) notes the importance of stories even further and expresses the importance of the story in an even more interesting way. Shattuck (1996) points out that :-

...... what I believe to be the best set of records we have about ourselves: stories of all kinds, true, embellished, invented.

We pass important elements of our culture on through history, which is our story, although we know accounts of it are seldom full or accurate. But also stories are an important tool in the enculturation of our children. The story of Cinderella exists in some 500 different languages.

The details of how the research was actually conducted need to be provided as do the particulars of the data analysis and the findings. Then the story needs to state what the significance of the findings is to the community of scholars and practitioners. And finally the story ends with some reflections on the limitations of the research and possible avenues for future research. The story has to be told in such a way that it reflects the fact that the researcher has acquired a deep understanding of the issues addressed by the research and that this has been acquired by the discourse between the researcher and the informants.

The story which relates all the major features of the case study is not always well told. Academic researchers are often too anxious to present the facts and the figures to do justice to a narrative which accounts for the context in which the research is being conducted and the nature of the variables which are being studied.

1.8. The scientific logic of the case study

The scientific logic which underpins the case study is closer to that of the experiment than that of the survey. Researchers distinguish between different types of experiment which include laboratory experiments, quasi-experiments and natural experiments. An experiment is sometimes described as a test or a demonstration which can be used to explore a theory or enquire into the operation of how a theory works. An experiment would also be entirely exploratory which could address a phenomenon without yet having any theory behind it. The logic of the experiment is that a researcher creates an environment in which he or she intervenes in order to see the affect of the intervention. This is the application of the principle of cause and effect. The experimenter compares the before and after situations and the changes which occur is the output of the experiment. In a laboratory experiment the researcher has a high degree of control and attempts to neutralise the affects of extraneous variables which may be present but are of no direct concern to the experiment. Control is an important issue and where there is a lower level of control the term quasi-experiment is sometimes used.

Natural experiments occur when the researcher has no control over the environment. What is termed a natural experiment is an unplanned, often natural occurrence, which is studied, together with the before and after situation, in order to establish the affect of the natural incident. There will always be more variables in the environment and the researcher has to be most careful that he or she understands which are the variables of importance. What is learned from a natural experiment does not necessarily lend itself to generalisation, but this does not detract from the value of knowledge established in this way.

The earthquake which produced the tsunami off the coast of Japan in 2011 and which caused serious damage to the Fukushima Daiichi Nuclear Power Plant complex could be considered by researchers to be a natural experiment as it provided an opportunity to study the after affect of a major nuclear accident. This natural experiment demonstrated just how vulnerable nuclear power plants can be and how difficult it is to

What constitutes Case Study Research?

provide support to the individuals involved in such an incident. The eruption of the Eyjafjallajökull volcano in Iceland in 2010 may also be seen as a natural experiment which highlighted how inadequate our knowledge of dust and airplane engine damage is. The essence of the natural experiment is that we are able to compare the before and after situation and to learn lessons from the affects of nature's intervention. Of course those people directly affected by the tsunami or the volcano's eruption, do not speak of these events as natural experiments, but as disasters.

In the world of the social scientist some radical change situations may be regarded as natural experiments. The deregulation of the Banking Sector and the financial success and then the financial collapses some years later may be regarded as a natural experiment. The election of a coalition government in the United Kingdom may also be regarded as such.

In order for something to be regarded as a natural experiment it is important to be confident that the principle *post hoc ergo propter hoc* applies. Post hoc ergo propter hoc is Latin for *after this therefore because of this* and this principle points out that because an event follows after another it is not necessarily because of the preceding event. This is a way of pointing out that the relationship of cause and effect is more complex than establishing the temporal order of events.

A social science example where there is some ambiguity about the post hoc ergo propter hoc principle is the drivers that caused the British Empire to be dismantled after the Second World War. Some commentators claim that it was the British fatigue with conflict and their near bankruptcy which caused the decline and the dismantling of the British Empire. Others would say that the fatigue and financial situation of the United Kingdom was not a direct issue, but that it was the awakening of the Empire's subject peoples to their right to freedom and self determination. Of course all of these claims may be true, but from a researcher's point of view it is important to be able to establish what the principal drivers and the principal effects were and there may be some debate about this. The relationship of cause and effect is often not as clear cut as researchers would like it to be.

The natural experiment calls on the researcher to carefully examine the before and after situation and to establish which aspect of the occurrence (i.e. in scientific terms the intervention) had a direct bearing on how the situation changed. This is similar to the logic of the case study.

In the natural sciences it is useful to learn from more than one incident and thus comparing the 2004 Indian Ocean tsunami to the 2011 Japanese tsunami may reveal more useful knowledge than studying only one such event. The same applies to comparing the Puyehue Volcano in Chile in 2011 with the Eyjafjallajökull volcano in Iceland in 2010.

Although the case study researcher is unable to exert control over the situation being studied, as would happen in an experiment, he or she examines the situation to determine which the most important elements at play are and what their consequences appear to be. In business and management, a case study will often be set around a new strategy, a new product range, the appointment of a new executive or some other such intervention.

What constitutes Case Study Research?

As in the natural sciences, multiple case studies may also be used in business and management studies. If a second case study is conducted then the researcher has evidence or data to compare and contrast and thus more learning may emerge. It is important to understand that a second, third or fourth case study is not conducted to replicate the first case study as would be done with experimental research and therefore there is no requirement to find a random or representative sample. Unfortunately this is not always made clear and some researchers argue that findings of case studies should be considered to be representative in some sense of a group of similar situations, organisations or individuals.

This is quite unlike the logic of the survey which is based on an attempt to find a result which will be representative of a population. Surveys require large samples which need to be selected on a random basis or at least be representative of the population being studied. Larger samples will not necessarily enrich our understanding of the situation being studied but they will generally increase the level of confidence with which we can claim our findings.

The term case study research is sometimes misused by those who do not know the requirements of the technique, but have heard the term and incorrectly believe that case study research is a matter of engaging in randomly selected multiple data collection activities, analysing them separately and then arguing that they constitute case study research. The result of this approach can be highly unsatisfactory. Fortunately this does not often succeed as supervisors or examiners will require the researcher to reclassify his or her research. As stated above, to be a competent research case study the nine issues listed need to be addressed.

1.9. Summary and conclusion

Competent case study research presents a rich multi-dimensional picture which is complex. There are a number of issues involved in the planning, execution and the analysis of a competent case study and for that reason a carefully prepared research protocol needs to be completed and approved before the research processes begin. It is not difficult for case study research to lose its focus and to drift away from its original purpose. Preparing a research protocol is time consuming and is sometimes resented by researchers who can think that research planning is a bureaucratic interference and time consuming overhead. Especially in case study research, this is not so. Careful planning is essential.

Case study research is an increasingly popular approach to research in business and management studies and it is something which every academic researcher in this field should be aware of and be able to evaluate if it is appropriate to use in his or her circumstances.

Chapter 2
Different Types of Case Studies and Choosing Appropriate Situations to Study

2.1. Different meanings of the term case study

The term case study is used in two different ways to mean quite different activities and it is necessary for the researcher to be clear on the meaning of the term case study in the context in which he or she encounters it. A case study may be a teaching-learning tool or it may be a research design and these two different usages of the term case study should not be confused.

2.2. Teaching – learning case study

From the point of view of those who are interested in teaching, the term case study is used to describe a story which could range from a single page up to perhaps a 50 or 60 page document. A situation is described in some detail for use in a classroom, to stimulate discussion which will lead to those present learning about the subject of the case study. This type of case study is sometimes described as a teaching – learning case study. The story told in these case studies may be real and possibly accurate or they may be fictitious. The objective of the case study is to challenge the readers and to provoke them into asking questions about the situation described as well as finding possible solutions to the problem articulated in the text. From this perspective case studies are often presented to a class which is invited to work in small groups to thoroughly discuss the issues presented in the case study. The small groups are then advised to reassemble or reconvene into the larger group for a full class discussion (Heath 2002). Sometimes case studies are presented to students with the additional complication of there being a considerable degree of time pressure. For example a 50 page case study may be presented to be read and discussed in a couple of hours. Only the fastest readers could comfortably cope with this. The pedagogical justification for this is the fact that some important organisational decisions are made under circumstances of pressure and without full information. This pedagogic approach is said to simulate real-life situations where executives are faced with complex problems, information overload and the need is to find solutions. Working in a group on a case study allows different people with different perceptions of the situation to compete in argument with others and thus the teaching – learning case study improves both the knowledge of the subject or topic and also the skills of discussion in competitive argument situations (Leenders M, et al., 2001).

As well as presenting this type of case study in a classroom some universities use case studies for examination purposes. In an examination situation case studies will normally not be as lengthy as they can be in a classroom situation because of the time required to read them, although some universities will actually provide examination candidates with the case study a few days before the examination and ask them to come to the examination having thoroughly read the docu-

Different Types of Case Studies and Choosing Appropriate Situations to Study

ment to face an examination which is composed of questions directly related to the case study.

Learning from case studies has a long tradition. Our knowledge of medicine for many years was largely based on examples of illness and injury which had been encountered on a case-by-case situation. It was only in modern times that there was any possibility of conducting effective medical research on groups. Herodotus (c 450 BCE) describes how the Persians used medical cases to help ill patients:

> *They have no physicians, but when a man is ill, they lay him in the public square, and the passers-by come up to him, and if they have ever had his disease themselves or have known any one [sic] who has suffered from it, they give him advice, recommending him to do whatever they found good in their own case, or in the case known to them; and no one is allowed to pass the sick man in silence without asking him what his ailment is.*

Even today, two and a half thousand years later, particular cases in medical research are important. As Kennedy (1979) remarked:

> *Studies of individual cases allow the evaluator to learn the intricate details of how the treatment is working rather than averaging the effect across the number of cases.*

The case study approach was also used in other fields of study. The practice of law and thus the legal profession is extensively influenced by case law. Although most countries have to a greater or lesser extent a constitution on which the laws are based, how the actual laws themselves are applied are often a result of cases which have come before the courts. Furthermore when legal decisions are handed down they are often conceived on the bases of precedent which is no more than saying that the judge has referred to previous cases.

The teaching – learning case study was famously developed by Harvard Business School and this approach has now been adopted in many universities around the world. Case studies written at Harvard Business School are sometimes presented in the form of a well written and engaging mystery story. The organisation, the characters involved and the processes used will be fully described but not the precise nature of the problems or issues which the students are required to address.

The idea of this is that by not supplying all the information the student will be required to create a list of missing facts. By not describing in full the problem the student has to work this out for him or herself. This is seen by Harvard as part of the learning experience. Some academics will tend to describe the type of case study produced by Harvard as a case history and it seems that this may well be a useful term to use. Another feature of these teaching – learning cases is that they can provoke curiosity and this is done to engage the student in the topic. Some of these teaching – learning case studies will be accompanied by a list of study and reflection questions, but sometimes they are not. The lack of questions is sometimes

Different Types of Case Studies and Choosing Appropriate Situations to Study

controversial. The justification for not supplying the questions is that the student should be sufficiently well versed in business principles to be able to determine what the important issues in this situation are and therefore be able to anticipate what the questions should be[2]. The reason that this may be contentious is that there are often multiple perspectives through which a student could look at a problem and in a teaching-learning situation the student may not have adequate time to give to a range of multiple perspectives. Although teaching – learning case studies are extensively used in business and management studies there are more than a few individual academics that do not hold them in high regard and consider them as over rated as a teaching technique.

2.3. The research case study

A research case study has a number of important differences to a teaching – learning case study. In the first instance the research case study needs to be an account of a situation in which there is a problem or a research opportunity. The case study needs to describe the context as well as the research activity and the result of this activity. This account of the research needs to be as accurate as possible. In addition and perhaps most importantly the research case study needs to be conducted in such a way that it answers the research questions. This is the purpose of the research case study technique. The research case study should be as well written and

---

[2] The term question here refers to the study questions which are associated with a case study used in the teaching-learning mode and not a research question. Research questions will be discussed later.

engaging as a teaching – learning case study but should not however be written as a mystery story.

2.4. Descriptive case studies or exploratory case studies

Before looking in some detail at the characteristics of case study research it is important to note that some researchers perceive case studies research as having different functions to others. Case study research is sometimes regarded as being a useful exercise in order to obtain background information about the intended research question and especially the context in which the research question will be studied. Often these case studies are referred to as descriptive case studies or exploratory case studies[3] and this type of research is generally not considered to be adequately formal to be able to make a 'serious' contribution to the body of theoretical knowledge. People who take this view will generally have a strongly positivist attitude to-

When is a single case study most acceptable?

1. If the case study is being used to test a well formulated and accepted theory.
2. If the case study is unique.
3. If the case is truly representative of a category of situations.
4. If the study is longitudinal with several data collection periods.
5. If access to a particularly suitable case study site will preclude the researcher from approaching other sites.

---

[3] Some academic researchers consider the use of a descriptive or exploratory case study an important step in arriving at an appropriate research problem and thus useful researcher questions.

Different Types of Case Studies and Choosing Appropriate Situations to Study

wards academic research and will expect to see more quantitative approaches used to justify the assertion that the results of the research are valid, reliable and generalisable.

On the other hand there are academic researchers who use case study research in a more in depth and formal way to obtain valid and reliable research findings. The issues of validity, reliability and generalisability[4] will be discussed later but for now it is important to say that if a case study has been used as part of a qualitative or interpretivist approach to research, it may be more appropriate to refer to the issues of credibility, transferability, dependability and usability than to validity, reliability and generalisability[5].

Besides the issues of descriptive case studies or exploratory case studies there are several other important different approaches to case study research and there are a number of different opinions as to how this type of research should be conducted. In Chapter 1 it has been stated that a case study will use different sources of data and even different research methods. One of the differences mentioned was that the case study could be positivist or interpretivist in orientation. As mentioned before when conducting a positivist orientated case study the researcher will be taking a deductive approach which will test an hypothesis or a group of hypotheses. The

---

[4] Sometimes the term replicatability is also used in this context. It has been omitted here because in most situations in social science research it is not possible to formally replicate a research study.

[5] Transferability and usability are the qualitative research equivalents of generalisability. Credibility corresponds to validity and dependability matches reliability.

researcher will be seeking quite specific data which he or she believes will directly address the claim made by the hypothesis. The data may be quantitative or qualitative but it will be quite specific. On the other hand if the case study is being used by an interpretivist researcher, an inductive approach will be used and the researcher will be seeking to establish new theory. There will seldom be a focus on finding specific data in the way that this is required when hypotheses testing is done. In general, inductive research has a much broader scope and may also be in more depth than deductive research.

2.5. A one case study research project

A question which is often raised is, "Can a doctoral degree be obtained on the basis of one case study or are a number of case studies required and if so how many cases?"

One case study

Those who argue for one case study sometimes assert that multiple case studies inevitably spend too much time and energy comparing and contrasting the situations in the different case study locations rather than focusing on the detailed nature of the matter they are studying.

In the first place it may be said that case study research is, in academic circles, perfectly respectable and the question of the number of cases required can often be easily answered. However the answer to this question is often dependent on the research question which is being addressed. The research question dictates many of the operational aspects of an academic research programme. It has already been stated that case study research is designed to answer complex or challenging questions. If the research question was sufficiently complex and if the case study led to a thoroughly profound under-

Different Types of Case Studies and Choosing Appropriate Situations to Study

standing of the situation and the production of a useful answer to the question, then it is possible that a one case situation could be adequate for a doctoral degree. What needs to be stated clearly is that it is not easy to achieve using one case study the level of understanding that is required so that examiners are sufficiently impressed to accept that a single case study is adequate for a doctoral degree. For this reason, a one case dissertation should not be encouraged for the purposes of a doctorate. There is also the fact that many supervisors do not want to supervise a situation in which there is only one case study and similarly it may be difficult to find an examiner who would be prepared to evaluate such a piece of work. However sometimes the situation is such that the researcher will only be able to obtain access to one organisation and then a single case study may become the only way forward.

2.6. More than one case study research project

Given that one case is not often enough then the question is how many cases are required? Again reference has to be made to the research question and to the intentions of the researcher. With an appropriate research question and with two particularly appropriate case studies it may be possible to produce a piece of work which is acceptable for a doctorate. Two or more observations i.e. case studies allow the researcher to indulge in some comparative analysis (using both similarities and differences) which can produce some useful insights into the nature of the circumstances being studied.

But in general two cases are often considered not to be sufficient. A doctoral degree candidate would be better advised to select three or four cases and sometimes perhaps even five. In academic research the cases are required to be in depth and more than five cases would be considered by many supervisors and examiners as more than enough work to undertake in the space of a 3 or 4 (full-time) to 6 (part-time) year doctoral degree.

With multiple case studies it is necessary to employ a multiple case study design with a technique for cross case analysis and the details of this will be discussed later.

2.7. The unit of study – the organisation is the unit of study

In doctoral research it is necessary to be clear on the unit of study, or unit of analysis, especially when it comes to using the case study approach. Often the whole case study is the unit of analysis; for example, a case study which is conducted at corporate level will address the organisation as a whole. This approach is referred to as a holistic[6] design. This design applies equally if a single case is being used or multiple cases are to be employed as is shown in Figure 2.1.

The left hand side of Figure 2.1 shows one case study addressing a major part if not the whole of the organisation. This is the single unit of analysis. The right hand side of Figure 2.1 shows a study with four different cases each of which addresses a case study as a whole part of an organisation.

---

[6] The term holistic was coined by Jan Smuts in 1926 in a book called *An enquiry into the whole*. Jan Smuts is better known today for his roles as a Field Marshal and Prime Minister of the Colony of South Africa than a leading world intellectual.

Different Types of Case Studies and Choosing Appropriate Situations to Study

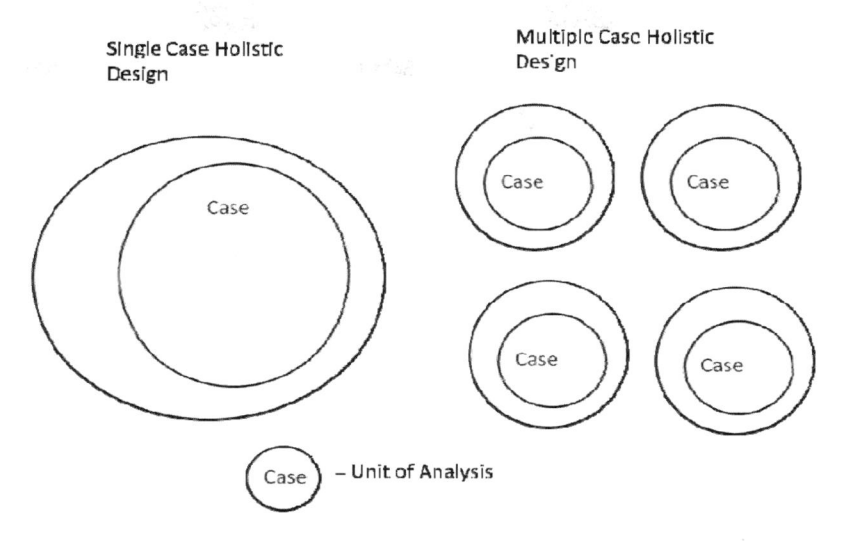

Figure 2.1: Single and multiple holistic case study designs

Working with multiple case studies offers interesting challenges. We normally work with multiple case studies in order to be able to obtain a fuller picture of what we are studying by having different situations to compare or contrast. For this reason it is necessary to ensure that our case studies are not incommensurable[7]. If for example we were interested in a research question related to Human Resource Management it might not be suitable to have a large global corporate entity and a civil service department, a public sector non-for-profit such as a museum and then a family business to compare.

---

[7] Bryman and Bell (2003) suggest that it would be possible to use multiple case study research to study a successful organisation and a poorly performing organisation. This could be the basis of a useful study provided the issue of incommensurability did not produce problems.

These cases might have so little in common as to make any comparison among them not useful if not actually meaningless. Of course there could be circumstances where such a comparison might be useful. The point is that care has to be taken in the choice of case study.

2.8. Multiple units of analysis

It is also possible to conduct a study which considers operating units within a larger organisation. In such an instance we can talk about there being embedded units within the case study. In this latter situation we have the opportunity for cross case analysis which allows comparisons to be made between the different units of analysis within the one greater case study environment. Keeping to a Human Resources Management theme it could be interesting to know how Human Resources Management Policy is implemented differently across the different divisions of a global organisation. We might want to compare and contrast policy implementation within the marketing function with that within the finance function or we might like to study an Asian operation and see how it compares with a Latin American operation.

This approach increases the potency of the case study approach as it delivers some of the advantages of the multiple case study design. The single case study with multiple embedded units is shown diagrammatically in Figure 2.2. The question of the number of embedded units which may be studied again related to the research question but in this situation there is an additional practical consideration which depends upon how many embedded units with which the researcher feels he or she can cope. Embedded case studies will consist of

Different Types of Case Studies and Choosing Appropriate Situations to Study

many variables and there will be a large number of informants and extensive background or contextual information to be gathered.

Single Case
Embedded Design

```
         ┌─────────────────────┐
         │        Case         │
         │  ┌──────────────┐   │
         │  │  Embedded    │   │
         │  │ Unit Analysis1│  │
         │  └──────────────┘   │
         │  ┌──────────────┐   │
         │  │  Embedded    │   │
         │  │ Unit Analysis2│  │
         │  └──────────────┘   │
         └─────────────────────┘
```

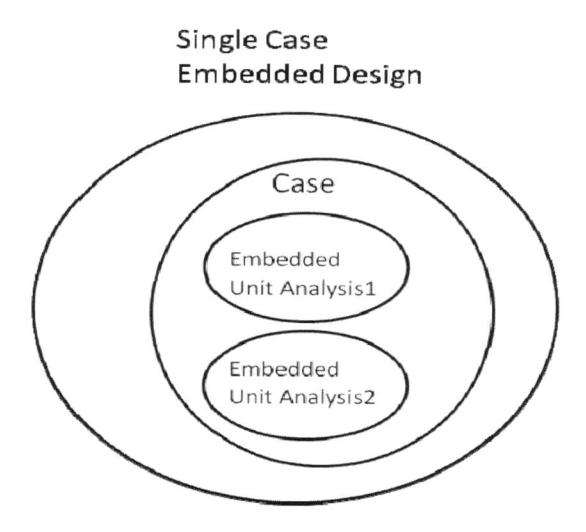

Figure 2.2: Single case with multiple embedded units

Researchers often find this research design attractive as it allows them to compare different units while keeping the greater corporate culture issues relatively constant. But it would be a mistake to suggest that all the corporate culture issues would be constant in such a research design. Some researchers will wish to examine a number of embedded units[8] but will also want to do so across a number of different case

---

[8] There are a substantial number of ways of tackling embedded case studies (Scholz and Tietje 2002) and those interested in this approach will need to spend time looking at the different alternatives available.

study situations. It is most important to keep a firm grip on the number of embedded unit and case studies as this type of work can easily become onerous.

Once again the number of units and case studies is related to the researcher's ability to handle a large amount of data. This type of study will create large volumes of data of various kinds. The design for this is shown in Figure 2.3.

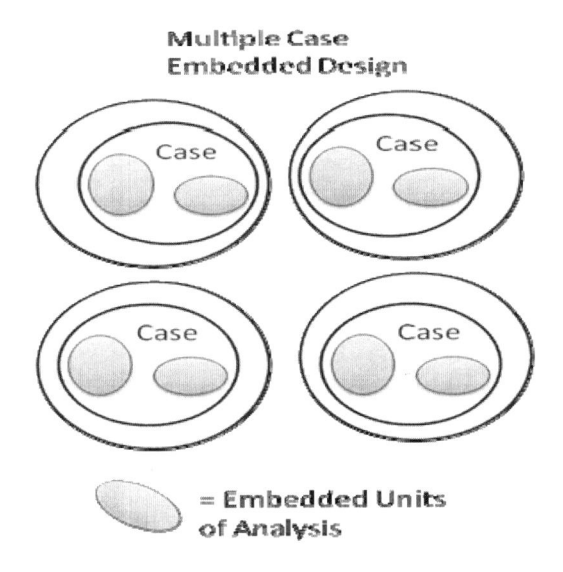

Figure 2.3: Multiple cases with multiple embedded units

2.9. Suitability and gatekeepers

Once again the number of case studies and the number of embedded units within each is a function of the research question but it really has to be emphasised that care needs to be taken not to overextend the researcher. The amount of

Different Types of Case Studies and Choosing Appropriate Situations to Study

work required to successfully complete a case study is often underestimated. The finding of suitable organisations with which to work; locating gatekeepers and then individual informants; capture of the data; creation of transcripts; analysis of the transcripts; writing up of the case study with the findings is more labour intensive than is often recognised. It is therefore important to keep the number of case studies and the number of embedded units under strict control, especially where doctoral research is involved and there is a time limit which should not be exceeded.

Moving on from the high level design issues, in general the choice of which organisations and which parts thereof to research is a difficult one.

2.10. Suitability Profile for a Case Study

As a lead into the case study protocol which is addressed in Chapter 4 it is useful to create a Suitability Profile for a Case Study. Such a profile will contain the requirements that a case study organisation should have in order to be included in the research. It is important not to waste time with organisations which do not meet the requirements as set out in the Suitability Profile for a Case Study.

> **Suitability Profile for a Case Study**
> Is the proposed Case Study organisation:
> 1. In an appropriate industry sector;
> 2. The right size by turnover, assets, market share or some other measure;
> 3. Sufficiently complex in nature to be interesting;
> 4. One where adequate access can be achieved;
> 5. One where they will allow their story to be told in an academic dissertation.
>
> Whatever the profile, for useful results a case study should be conducted in an organisation which is:
> 1. Relevant from the point of view of the research question;
> 2. Significant in terms of having something important to say;
> 3. Accessible in terms of geographic location;
> 4. Amenable to the study in terms of staff cooperation.

If any one of these five factors in the adjacent box is missing from a proposed case study location then it should be omitted from the research frame.

The research question is the fulcrum around which all other academic research activity takes place. If a proposed case study is not directly relevant to answering the research question then it is unlikely to deliver much value to the research and should be passed over.

Only organisations which are significant[9] should be studied. Thus organisations which are too small or maybe even too large may not be suitable. Organisations in different countries or different cultures may also not produce data which would make a significant contribution to answering the research question.

It is important that the re-

---

[9] The word 'significant' here does not imply any statistical significance. Here the word means important enough to be included in the research.

Different Types of Case Studies and Choosing Appropriate Situations to Study

searcher be realistic in terms of the distance which may have to be travelled to conduct a case study. Minimising travel should be a priority of the research as it saves time and money and if the case study location is local then multiple visits to the case study site/s are much easier. It is unlikely that all the data required for a case study will be acquired with only one visit to the case study location. Sometimes three, four or five visits are required.

Researchers need to find a gatekeeper who will introduce them to the necessary people in the organisation. There can be several gatekeepers to the same organisation and sometimes it is useful to identify and acquire the cooperation of multiple gatekeepers. The researcher needs to find as many senior people as possible to take this role. If there are multiple gatekeepers then where possible the original gatekeeper should be asked to introduce the researcher to the others. But even with the backing of a senior gatekeeper there may well be informants to whom the researcher may wish to talk who do not want to be involved with the research. The researcher needs to respect this. By the way researchers will sometimes find gatekeepers who are antagonistic towards them. Personal assistants to senior managers or directors will sometimes try to protect the time of their colleagues by discouraging the researcher from seeking interviews[10].

---

[10] Well established researchers will recount stories of how to get around such situations. One way is to accept an appointment with another member of staff and when on the premises of the case study call on the original preferred informant informally to say hello and ask for an appointment.

31

2.11. Summary and conclusion

It requires careful thought on the part of the researcher before the choice of case study research should be made. One of the important considerations which has to be addressed is the possibility of obtaining adequate access to useful data through relevant and significant case study sites. This is often a major challenge. The difficulty of access to informants in order to obtain appropriate data should not be underestimated. If appropriate gatekeepers cannot be found then it may be necessary to find other case study sites or even abandon the idea of a case study altogether.

Chapter 3
Data: Quantitative and Qualitative

3.1. The importance of data

Being clear about the data requirements for the research and how to find and use appropriate data is especially important for case study researchers. It is more challenging in the case study context because there are so many different data sources available. It is worth saying immediately that case study research, perhaps more than any other type of research, will extensively rely on both primary and secondary[11] data. Primary data may be collected directly from or through people involved in the case study location[12] and secondary data will be obtained from previously published sources which describe the case study situation.

3.2. A definition of data

Although data plays a central role in research little attention is given to the definition of data. There seems to be an unspoken agreement concerning the lack of need for a definition of this word. When a definition is required then the model articulated by Russell Ackoff (1989) which describes data as part of a

---

[11] There are substantial secondary data resources available on the web and the use of this has been referred to by some researchers as e-Research.

[12] The term location used in this context refers to the organisation which is being studied rather than one physical site.

chain which starts with data, then moves on to information and then to knowledge, before arriving at wisdom, which is usually provided. However the Ackoff model is not particularly useful especially to researchers. If the enquirer presses for more information about the nature of data then it is often said that data is raw. An example of how obtuse or opaque a definition of data can be is given by Bellinger et al. (2004) when they wrote:

> *Data is raw. It simply exists and has no significance beyond its existence (in and of itself). It can exist in any form, usable or not. It does not have meaning of itself. In computer parlance, a spreadsheet generally starts out by holding data.*

This description is of little value to anyone and it certainly does not help an academic researcher. The term 'raw' is interesting, although it is left unexplained. It does conjure up the idea that data could be 'cooked'! More realistically the word 'raw' suggests that the data is independent of the person who has collected it and it is difficult to see how this could be the case especially in the world of academic research. A more insightful or reflective approach to understanding data is required.

Data normally occurs as a result of some phenomenon or event taking place. If the stock market is trading then the data from the trading represents all the buying and selling which took place that day. If the stock market was closed and there was no trading then the recording of no sales would also be data in its own right. This is an important feature of data which is that a non-event can also be data. In literature this is

Data: Quantitative and Qualitative

most strikingly described by Sherlock Holmes (Conan Doyle 1892) in the novel Silver Blade when he remarks about the dog that was not barking in the night. In this Sherlock Holmes story the lack of barking is a clue that the thief who struck at night was known to the dog.

> **Data and the research question**
>
> Except for some background and context data the primary data activities need to use only that data which is directly related to understanding and answering the research question. It is all too easy to lose focus and this needs to be prevented where possible.

Data is perceived by the researcher and it may be as a result of any activity which causes the researcher to receive a sense stimulus. Thus hearing an informant in an interview may result in data, seeing an eclipse may result in data, smelling a sewer may result in data, hearing the Great Bell of Bow Church (St Mary-le-Bow) ring out may result in data and so on. The expression "may result in data" is used because the researcher has to believe that the stimulus perceived will lead to his or her being able to better understand and answer the research question before it is recognised as data. Sounds, sights, smells and so on which contribute nothing to answering the research question can be regarded as 'noise'. In this context noise is any stimulus which interferes with better understanding and answering of the research question. There is a problem however which is that there are no universally agreed rules for determining what is and what is not noise. The decision to classify something as data or as noise is made by the researcher and he or she may be wrong. In everyday life doctors are known to misidentify symptoms, which can lead to grave medical problems which often amount to the same problem of not being able to recognise valid data as opposed to eliminating noise.

3.3. Do, dare, dedi, datum

Before examining the many issues related to data it is important to comment on the word data itself. Data is regarded by many to be the plural of datum. Of course to take this view one has to have some knowledge of Latin and not all English speakers have had an education which included Latin[13]. Datum is a part of the Latin verb which is *do, dare, dedi and datum* and this verb translates into English as to give. Some researchers argue that the word data, consciously or unconsciously, is too strong for what is generally found in research i.e. to say something is given may be taken to mean that it is correct and that it is unquestionable. This is worrying because the assumptions underpinning the data may not have been considered or questioned adequately. If this idea is accepted and thus the word data is not freely used then the word evidence may be a better choice. Evidence can be used with the notion of suggestion. Thus one can say, *The evidence suggests that*

> **The old debate about quantitative versus qualitative**
>
> It is increasingly accepted that there are very few circumstances where research will be only quantitative or qualitative. Increasingly it is being understood that most research requires evidence from both of these sources.
>
> It is time that the quantitative and qualitative warriors gave up quibbling and came to terms with the reality of 21st century research.

---

[13] Latin is no longer routinely taught in English speaking schools, it is however still taught in some expensive private schools, it's passing has been lamented by some academics.

Data: Quantitative and Qualitative

the marketing group has performed better than expected, for example[14]. Another word related to data which may also cause some concern is 'fact[15]'. Although we have an aphorism which declares that *'Facts speak for themselves'* this is seldom if ever correct from the point of view of the academic researcher. In addition some people talk about different types of facts. There are disputed facts and undisputed facts. In between these two positions there are consensual facts, probable facts and possible facts[16] (Bannister 2005).

Facts, data and evidence have to be interpreted by the researcher. Nothing should be taken at face value without careful consideration of the assumptions on which data, facts and even evidence are based.

3.4. Unit of analysis

In determining what data may be required an important consideration is the unit of analysis. If the research addresses the detail of consumer behaviour, such as how individuals choose products in a supermarket, then the unit of analysis of analysis will primarily be the individual. There may also be a secondary unit of analysis which in this case would probably be the supermarket. However if the research focuses on how organisations decided to allocate resources in global markets then the

---

[14] Evidence is a better word to use in qualitative academic research but unfortunately nearly all for the hundreds of textbooks and papers written on the subject of the analysis and interpretation uses the word data.

[15] The word 'fact' is derived from the Latin verb to make and also produces issues for the careful researcher.

[16] This taxonomy of facts is not universally recognised by scholars, but it is an interesting attempt to make a distinction between different types of facts for which some scholars might find a use.

unit of analysis might be national capital markets and perhaps shareholders' expectations.

The unit of analysis is important to establish early on as it will strongly influence the type of data required for the research.

In some cases the unit of analysis could be an event. Examples of events would be the launch of a new product, the insolvency of an organisation or the appointment of a new chairman of the board of directors. It is thought that case study research lends itself to working with events which often require multiple sources of data to acquire an understanding of the situation.

3.5. Quantitative and qualitative data

There are two major classifications of data which are quantitative and qualitative data and it is necessary to explore both of these in some depth.

Regarding qualitative data, although data exists when the researcher hears the informant during an interview, it normally only acquires a 'tangible' or at least an operational form which is of direct use to the researcher, when it is reduced to text. In practice the researcher will often record an interview and at a subsequent date he or she will transcribe the recording onto paper. It is this transcript of the interview which the researcher will normally work with and which he or she will consider 'the data'. The process of transcription clearly indicates that the data obtained in the above way cannot be considered as 'raw'.

Data: Quantitative and Qualitative

Data exists independently of the phenomenon, event or object which it describes. In that sense data is an abstraction and rather obviously it is not the phenomenon, event or object itself and as with all abstractions it is not the same as the event itself. Such an abstraction is created, as is the case with all abstractions, in someone's mind and the way it is created is a function of the knowledge, understanding and values of the person creating it. No two abstractions based on the same event can be guaranteed to be the same. In other words the perceptions of any two people will not be quite the same. Kuhn (1970) explained this issue when he wrote:-

> *If two people stand at the same place and gaze in the same direction, we must, under pain of solipsism, conclude that they receive closely similar stimuli. But people do not see stimuli; our knowledge of them is highly theoretical and abstract. Instead they have sensations, and we are under no compulsion to suppose that the sensations of our two viewers are the same... Among the few things that we know about it with assurance are: that very different stimuli can produce the same sensations; that the same stimulus can produce very different sensations; and, finally, that the route from stimuli to sensation is in part conditioned by education.*

Quantitative data is regarded as being less reliant on the knowledge, understanding and values of the person 'creating' it and thus is regarded as being objective. However there is a range of quantitative data and some forms of data that are treated as quantitative data are less objective. At the one end of the spectrum quantitative data could constitute measurements of distances between cities or houses or desks in an office. In fact quantitative data is the result of a magnitude assessment of any measurable dimension of a phenomenon or event. Speed of motor vehicles, bonuses of senior executives, balances in bank accounts, age of paintings of the masters, how many dollars are paid in bonuses to investment bank employees, weight of aircraft, number of widgets in inventory, percentage of market share, wages per hour are just a few such quantitative dimensions.

There are different categories of quantitative data which are referred to as categorical data, ordinal data, interval data and ratio data. Each of these different types of data are amenable to different degrees of summarisation and analysis with categorical data having the least scope and ratio data having the most potential for analysis.

Provided there is agreement on the measuring instrument and that it is a stable, reliable device then this type of measurement will generally be objective. However there can be disagreement about the appropriate units of measurement, the accuracy required and a variety of other matters which could produce different results. Nonetheless it is relatively easy to establish that if one desk in an office is five feet from the door and another is six feet from a door then the second desk is further away from the door than the first desk and the difference is one foot and everyone will understand what this means.

In business and management research attempts are made to measure opinions, beliefs and other non-tangible aspects of human experience. The measurement of satisfaction is one such variable or construct which is frequently subjected to assessment. Using Servqual, which is an established approach to measuring satisfaction, it is possible to say that with regards to one service issue the satisfaction level is 3 whereas for another issue it is 4. However we cannot say that everyone will agree on the meaning of the one unit of difference between these two 'measures'.

3.6. Different realities or different lens

It is sometimes argued that different people have different realities and therefore they will perceive any situation differently. Such a statement may be largely true but it is not particularly helpful. The word reality is sometimes challenged and it has been suggested that a better description is that everyone has a unique lens through which they view and understand the world around them and the uniqueness of each person's lens means that they will have different perceptions and understandings of what is happening. Although it is sometimes possible for the different perceptions and understandings to be substantial, often there is an adequate degree of common understanding for people to be able to create abstractions which may be used by others. Academic research normally works on the basis of this assumption.

Another feature of data in the research context is that when it is analysed it is often historic and as such it is not possible to recreate the circumstances precisely in which it was first obtained. In social science this is more obvious than in other sciences but at least to some extent it applies to most scientific explorations. We cannot recapture the same data because we have changed by the capturing of the first data set. We cannot observe the same people because they will have changed by the first observation and if we use different people then we are not repeating what we did before.

The ending of the Robert Louis Stephenson (1886) story of Dr Jekyll and Mr Hyde is based on this idea. Having run out of the ingredients for his potion which allows him to transform himself from the genteel Dr Jekyll into the anti-social Mr Hyde, Dr Jekyll sends out his butler to find and purchase some more of the requisite chemicals to facilitate this transformation. The chemicals are duly acquired and presented to Dr Jekyll. However despite following this own procedure precisely, this new batch of chemicals does not produce the same effect as was achieved previously. The reader is then invited to reflect on whether there had been some impurities in the first batch of chemicals which caused the monstrous transformation affect. Of course, there could now be impurities in the second batch of chemicals which would also cause the transformation to fail. Robert Louis Stephenson leaves the reader to contemplate why the procedure no longer works. It could also be that Dr Jekyll's physiology could no longer tolerate the transformation and he became immune to the potion. Whatever the cause, at this point Dr Jekyll was not able to repeat what he had done before. Much of social science cannot be replicated and as such it has the same problem as Dr Jekyll when it comes to repeatability.

After the decision regarding the research question, the choice of what data to acquire is perhaps the most important matter to be settled for any academic research project. Poor decisions related to data can cause much extra work to be necessary and in the worst instances can cause the research project to completely fail. Consequently the question of determining

Data: Quantitative and Qualitative

the appropriate data which will be acquired and the process of acquiring it, is of considerable importance.

3.7. Natural and elicited data

Both quantitative and qualitative data may be natural or it may be elicited. Natural data could consist of a list of the ages of senior executives in a particular company which is supplied from already established records. It could consist of the total payroll for each period. It could be the total of the amount of money on the till slip of employees when they buy their lunch in the canteen at work. Elicited data consists of data that is only revealed by the answering of a question put to the informant by a researcher. Thus when a researcher asks an informant to declare the extent that he or she agrees with a statement such as, *I regard the customer service of this Bank to be excellent, score 1 for complete disagreement and 10 for complete agreement,* the answer to this question may be regarded as elicited data.

It is thought that natural data has a greater degree of authenticity and perhaps less bias than elicited data as natural data is a matter of record while elicited data is subject to the state of mind of the informant when the question is being asked and this can vary from day to day if not from hour to hour or moment to moment. Also the chemistry between the researcher and the informant can produce different results when elicited data is required. The other issue with regard to elicited data is that, despite pre-testing and field testing, it is not always possible to ensure that each informant understands the question in the same way which according to Silverman (1995) is critical. Unfortunately it is not possible to be certain about how each informant actually understands any given question. But this is not as much of an obstacle as one might think as academic researchers know that certainty has little place in this type of research.

3.8. Hard and soft data

Quantitative data is sometimes referred to as *hard data*. This is because in some circles quantitative data or numbers is seen as more reliable and less subject to bias than qualitative data. By those who make this argument, qualitative data is sometime called *soft data* which is the opposite to *hard data*. But, this type of argument is not strong. There is no reason why numbers are intrinsically more reliable or less biased than words or images etc. Paulos (1998) points out that numbers are often based on assumptions which when closely inspected may not be as sound as they first appeared. Some researchers have said that the only number on the balance sheet of an organisation which is not based on an assumption is the line which represents the cash in the bank. However this is only true in simple cases. If there are multiple currencies involved then the exchange rate between them changes on a moment by moment basis so even for the cash at the bank there are assumptions made which may not be accurate. Many experienced researchers argue that the distinction between *hard* and *soft* data is not helpful.

In addition there is an increasing move towards the use of both quantitative and qualitative data in the same study. As mentioned before qualitative data is required before a quantitative measuring instrument is developed and qualitative data underpins any final interpretation of any research results.

Data: Quantitative and Qualitative

3.9. Data can be overwhelming – data overload

Researchers can be inundated with quantitative data. Statistical or mathematical techniques are used to assist the summarisation and analysis of the data and this may be conducted descriptively or inferentially. In both cases these techniques are used to facilitate the greater understanding of the data by seeking to establish patterns[17] in the processes being studied. Sometimes these patterns are studied deductively through the use of hypotheses testing or they can be studied inductively by seeking new theoretical explanations. Academic research will normally require as high a level of sophistication of summarisation and analysis as the data will allow. It is for this reason that statistics plays an important role in research and is well expressed by Agresti and Franklin (2007) when they wrote:-

> *Statistics is the art and science of designing studies and analysing the data that studies produce. Its ultimate goal is translating data into knowledge and understanding of the world around us. In short statistics is the art and science of learning from data.*

But statistics is not the only way of learning from data. Research has today moved a long way beyond needing a data set of some sort of numeric measurements before it can be authentic. Data may have many different stories to tell besides the learning which may be acquired from what would be re-

---

[17] Michael Shermer has coined the term 'patternicity' to describe the human mind's propensity to find patterns http://www.youtube.com/watch?v=Jv4KtAnjqzU Shermer points out that patternicity can produce results which are not valid and researchers have to be on their guard to prevent this from occurring.

garded as traditional quantitative analysis. Kennedy (1979) was aware of this when she said:-

> *It is important to realise that non-statistical arguments need not be invalid. Yet many researchers may be timid about attempting such inferences simply because the rules as to what constitutes a reasonably sound inference are ambiguous, relative to the rules as to what constitutes a sound statistical inference. What is needed are rules of inference that reasonable people can agree on.*

Although Kennedy wrote the above more than 30 years ago there are still a number of issues related to the validity of non-statistical arguments which are at least in part due to the large number of different approaches to the collection and the analysis of qualitative data.

As previously mentioned qualitative data is varied in nature and may be acquired from a range of sources. Insightful qualitative data can be difficult to access and demanding to work with. Qualitative data is so ubiquitous and so varied that it is difficult to define but the concept of qualitative research is well established.

According to Strauss and Corbin (1998):-

> *By the term 'qualitative research' we mean any type of research that produces findings not arrived at by statistical procedures or other means of quantification.*

Although this definition is accurate it is unfortunately a negative one. The problem is that almost every aspect of our ex-

Data: Quantitative and Qualitative

periences can lead to qualitative data and thus it is a challenge to formulate a concise definition.

It is therefore better to describe the characteristics of qualitative data as follows:

1. Qualitative data may be gathered/captured/collected during many different sense experiences including observation of people in their daily lives, conversations - formal and informal, feeling temperatures or the texture of materials, registering olfactory sensations or through tasting experiences;
2. The researcher's knowledge and prior experience determines whether some stimuli should be regarded as data;
3. In obtaining qualitative data there will be a high degree of involvement between the researcher and the data/evidence;
4. Qualitative data/evidence collection will be flexible and sensitive to the context in which the research is working;
5. The researcher will be the arbiter of whether a stimulus is data or noise.

From the above it is may be deduced that a researcher, especially one seeking qualitative data, needs to be sensitised to that data and where it may be found. Before exploring the issue of sensitisation to data it is useful to consider the primary sources of qualitative data in business and management study research projects.

3.10. Commonly available sources of qualitative data

The following is a list of the more commonly available sources of qualitative data.

- Observation of live situations
- Reviewing photographic evidence and video data
- Interviewing including listening to groups such as focus groups
- Field notes
- Ethnographic fieldwork including participant observer work
- Archived texts
- Physical artefacts
- Data caches such as diaries, personal correspondence

> **The importance of data**
>
> It is difficult to exaggerate the importance of data. Data primary or secondary is always needed in academic research. But it always needs to be remembered that with case study data the researchers is often obtaining the interpretations of informants. In fact what is actually happening is that the researcher is working his or her interpretations of the informants' interpretations. Without highly honed listening skills and an appropriately tuned imagination the type of data so acquired can have a reduced value.

Observation of live situations is an important source of qualitative data. These observations may include human interactions with other humans. They may also involve human interactions with machines or human reactions to organisational procedures to mention only a few instances where observation may play an important role.

Reviewing photographic evidence and video data is another potentially rich source of qualitative data. The photographic evidence and video data being

Data: Quantitative and Qualitative

reviewed may have been created by the researcher in observation and in this case the photographic evidence and video evidence is being used as an aide memoire. On the other hand there are extensive libraries of photographs and videos which may be accessed and which can contain rich data and these are perfectly acceptable sources for qualitative research.

Interviewing including listening to groups such as focus groups is one of the most popular ways in which qualitative data is sourced. In general, interviewing is a one-to-one activity where the researcher is soliciting data from an informant. Sometimes the informant may supply the researcher with numeric data which supports the informant's views. But the interview is quintessentially a qualitative data collection activity. Focus groups are often perceived as a variant type of interview. Although this is not strictly correct from a data categorisation point of view, focus groups normally supply qualitative data.

Field notes are memorandums or aides memoires made by the researchers of their own impressions of events, processes or people they encounter during their field work.

Ethnographic fieldwork including participant observer work will lead to the collection or creation of qualitative data. Ethnographic fieldwork including participant observer work involves the researcher participating in some aspect of the daily work of the organisation and the people being studied. When a protracted period is involved and the researcher is able to claim that he or she has had a lived experience with the organisation and the people being researched, the research is referred to as ethnographic. Max Travers (2001) describes the approach to ethnography as requiring:

'an intimate familiarity with that experience and the scene of its operation'.

Further he asserts what is required is a 'human lived experience'. The quality and the precision of the account of the experience is a central issue in undertaking ethnographic research.

When a shorter exposure is involved and only a relatively superficial knowledge is achieved the researcher is referred to as a participant observer.

Archived texts refer to documents which range from formal published financial accounts, operating manuals, advertising and publicity material, to e-mails and private correspondence. These can be current texts or they could be documents which have been created and/or used in the past.

Physical artefacts refer to any tangible objects which can provide evidence which may be helpful in answering the research question. This could be an elderly out-of-date computer in the accounts office. It could also be an oversized antique oak desk in the managing director's office. Different products sold by the organisation over time are good examples of physical artefacts that can be really important to a researcher. Uniforms worn by staff at different periods can give an impression of how the organisation viewed its staff.

Data caches is a term used to describe all other sources of qualitative data. Examples of these are scrap books containing

Data: Quantitative and Qualitative

press releases, diaries of appropriate individuals, copies of speeches, photographs and video recordings, public and private databases, messages pinned to notice boards, and passages from literature. In his novel *Hard Times*, Charles Dickens provides useful information about the attitude of wealthy to the working classes and especially to working class children and thus it may be regarded as being a part of a data cache for appropriate research questions.

Both the quality and the quantity of data obtained from these sources is a function of the skill of the researcher. Highly skilled researchers will be able to find the necessary data required for their research while newcomers to research often struggle. One of the essential attributes of a competent researcher is that he or she needs to know what to look for. This is referred to as being sensitised to both the context of the research and the implication of the research question. Sensitising a researcher is not simple and there is no one formula for so doing. Brainstorming sessions with a supervisor may help. Periods of reflection are also useful. If the researcher is not adequately sensitised then valuable data may be overlooked.

In some research circles qualitative data and qualitative research have been regarded disparagingly. It is not simply a convenient escape route for researchers who do not wish to employ statistical analysis. Using qualitative data is not an easy option. It is time consuming, labour intensive and subject to misunderstandings. In short, often much harder work is required with qualitative data than with quantitative data. But it is still the case that those who are committed to positivistic research methods sometimes attempt to question the validity of qualitative data and research. Fortunately there are an increasing number of competent and articulate researchers who can explain the value of qualitative data and research and why its detractors are ill informed.

When it comes to the question of data case study researchers will normally cast a wide net and thus have at their disposal multiple sources of data. Although this can be very enlightening and stimulating for the researcher, it is also a substantial challenge. It is by no means obvious which sources of data will afford the best insight to understanding and answering the research question.

In general not enough thought is given to the data requirements for academic research projects. Researchers often rush the data decision without adequate reflection as to what data is required and how they will be able to obtain it. It is useful to have a Data Acquisition Plan. The form of this plan requires the researcher to consider what data is required to answer each aspect of the research question and to think about how the data will be acquired and from whom it will be required. It is simply another useful check list.

3.11. Combining quantitative and qualitative data

One of the important features of a case study is that it allows multiple sources of data to be used. This may include quantitative and qualitative data. Where this is the case the two different data types need to be analysed and interpreted separately and then the researcher has to weave the results or the analysis of the results of the quantitative and the qualitative data into one argument. It is the creation of a convincing argument which is central to research success. It is not trivial to

Data: Quantitative and Qualitative

create such an argument and the researcher needs to devote a considerable amount of time and energy to this. When the quantitative findings of the research support the qualitative findings then the construction of the argument is relatively straightforward. However sometimes the quantitative findings and the qualitative findings point in different directions and when this happens the researcher is faced with a much greater challenge. This does not mean that the research is fundamentally flawed but rather that more thought needs to be given to some of the assumptions underpinning the whole project.

3.12. Data Acquisition Plan

As data is the basis of discovery in empirical academic research its acquisition should be planned from an early stage and Figure 3.4 suggests a matrix which is useful to consider during the early planning stages of a research project.

Research questions	Data required	Organisation/s	Informant type	Remarks – Analysis Type
Main research question				
Research sub-question 1				
Research sub-question 2				
Research sub-question 3				
Research sub-question 4				
Research sub-question N				

Figure 3.4: Data Acquisition Plan

In empirical research projects not planning the data capture can be a catastrophic mistake. Access to data is critical and it is often quite difficult to obtain. Organisations and even individuals can be quite reluctant to allow academic researchers to read reports and speak to key individuals. The most frequently cited reason for this is that of confidentiality and the fact that information about the organisation can have commercial value. However, it has been suggested that the real reason is often that management feels that if facts and figures are provided they will in some sense be vulnerable to direct or indirect criticism. It is not easy to counter this view as it is to a large extent correct. Academic researchers are expected to show the skills of critique and as such they will often ask penetrating questions which management can find uncomfortable or even threatening. This is a difficult problem because a research that appears in any way threatening is unlikely to last any length of time as the guest of the organisation. This is particularly true if the researcher is a young PhD degree candidate.

3.13. Summary and conclusions

Being aware of the different types of data classification and the different sources of data are important aspects of planning any academic research project. In many research situations both quantitative and qualitative data are required. There are many sources of data and the researcher needs to be sensitised to these so that he or she can take advantage of the wide range of data available.

In general qualitative researchers are often more aware of this despite the fact that quantitative researchers especially need to be able to understand the context of the numbers they use.

Data: Quantitative and Qualitative

Numbers without context are inevitably of no value. This is expressed by Paulos (1998) as follows:-

Without an ambient story, background knowledge, and some indication of the providence of the statistics, it is impossible to evaluate their validity. Common sense and informal logic are as essential to the task as an understanding of the formal statistical notions.

Case Study Research

Chapter 4
The research proposal and protocol

4.1. Understanding and planning academic research

All research benefits from planning and structure. Randomness does not facilitate discovery, although it is not impossible that knowledge could be created in an unplanned way and the history of science does suggest that this has happened on occasions. It is possible, after all, for a person to walk down the street and have the wind blow a piece of paper into his or her face, which turns out on inspection to be a winning lottery ticket. Although this is not impossible it is highly improbable to the point of not being a useful way of think about how to win a lottery.

Taking a planned and structured approach to research does not over restrict either the research or the researcher. It does not mean that the researcher needs to be inflexible. Researchers should always be able to respond to the events which take place during the research process. It is not possible to foresee in any detail how a research project will unfold and if the attitude of the researcher is that the original plan has to be followed at all costs then the research will often fail to meet its objectives. The proper approach to research planning is reminiscent of Dwight D Eisenhower's remark which was, *'Plans are nothing; Planning is everything'*. The other comment on planning for which General Eisenhower is famous is *'Plans are great until you encounter the enemy'*.

There are however some academics who strongly oppose the idea of planning a research project. These academics argue that research is always a voyage of discovery and that it is actually impossible to know in advance where this journey will take the researcher. It is said by such researchers that the word protocol indicates a positivist attitude to the research. The recommendation is that a softer word such as research blue print may be used to describe whatever 'forward thinking' is done and this 'forward thinking' should be seen not as a course of action but as an attitude towards what has to be done during the research. This approach can be referred to as emergent planning and those who disapprove of it may sometimes label it as a postmodern approach to research. This approach does not suit many researchers and is considered by most to be challenging. This is mentioned for completeness and it is not recommended in this book.

4.2.   The need for flexibility

The degree of flexibility required with regard to an academic research project varies enormously. Sometimes it will be a matter of changing the thinking about the number of informants required. On other occasions it will be a question of changing the interview schedule or perhaps changing the software package with which the analysis will be done. But sometimes the change required can be far more fundamental and the researcher may, having followed the original plan for some months, decide that the process needs to be rethought from fundamentals upwards. Of course, if this can be avoided the research will have a much better chance of being concluded in time, on budget, and generally being more satisfactory to the degree candidate and the supervisor. But this will not always be the case if there is a serious flaw in the logic of the original plan and researchers should be open to the possibility of major rethinking.

Many academic researchers at doctoral level are surprised at how much work there is actually required to complete the degree. There is a considerable amount of academic or intellectual effort required which needs to be managed. But in addition there is considerable non-intellectual work. Assuming there is a great library facility supporting the work so that books and journals do not have to be fetched and carried, there is searching for references, sorting out those that are appropriate, and the creation of a catalogue of references. Having established a research instrument, either questionnaire or interview schedule, then there is the finding of appropriate informants. This can take long hours of phone work or even visiting the premises of prospective informants. If questionnaires are used then there may be a need to distribute them or it could be necessary to have them uploaded onto a suitable website. This requires learning a software package. These are just a few of the work activities which may be required when working towards a research degree. There are many more. It is hoped that this short list will indicate that an academic research degree is a major project and that its planning deserves attention.

4.3. Organisation and planning

Case study research requires a high degree of organisation and planning on the part of the researcher and this should begin with the preparation of a Research Proposal. The Research

Proposal takes the form of a document in which a prospective research degree candidate demonstrates that he or she is familiar with many of the academic issues which need to be addressed during the research degree they propose to undertake. This is achieved by a statement of the problem which underpins the research question and a discussion of the methodological choices available for answering the question. Both these discussions need to be illustrated by reference to the extant literature.

The Research Proposal allows the prospective research degree candidate to demonstrate to the university or business school that he or she is not naive about the topic to be studied or the methodology which is being proposed. The research proposal will also show to what extent the prospective research degree candidate is able to write academically.

4.4. The Research Proposal

The length of the Research Proposal varies considerably from university to university, school to school and so on. Some universities require a short research proposal but others demand a much longer document. The range is between 10 pages and 100 pages. The justification for the 100 page document is that if a prospective research degree candidate is prepared to put this amount of work into a proposal there is a considerable likelihood that he or she will complete the degree.

Some universities will expect a research proposal to be completed within a few months, say three to six months, while others will give prospective research degree candidates a year or even more. The research proposal will often be used to trigger the formal registration of the degree candidate. Before

The research proposal and protocol

the research proposal is submitted and approved the student may be in some sort of transitionary position. Some universities make a distinction between enrolment and registration. Such institutions will typically say that the individual may enrol but may not be registered until the research degree proposal has been accepted. Other universities may use a research proposal as a device to allow the transfer of a prospective research degree candidate from a master's register to a doctoral register. Sometimes the terms provisional and final proposal are used with a final proposal being required to obtain full registration for the degree. Although this may vary considerably from university to university the format or layout of a research proposal will follow to some extent the format of the final dissertation itself and will address the issues shown in Table 4.1.

Table 4.1: The layout of a research proposal

No.	Title	Content	Pages
	Cover page	Comply with the requirements of the Faculty or the School.	
1	Introduction	Point out the nature of the topic, its importance and describe the format of the rest of the study.	5
2	Literature review and research question	Demonstrate sufficient knowledge of the extant literature to understand what the research involves and what will be needed to answer the research question.	7
3	Research design	Address some of the research methodology options available and indicate the research degree candidate's preferences with regards to research methods. This section will usually make reference to the research ethics issues.	7
4	Executing the research	Indicate that the challenges which will present themselves have been given appropriate thought.	3
5	Possible findings and conclusions	Some indication of what the research might produce.	3

| 6 | Possible limitations | An awareness of the limitations of academic research at this level given the time scale in which it has to be completed and the funding available. | 3 |
| | Appendices | A list of possible appendices | 1 |

The research proposal outline shown in Table 4.1 amounts to 29 pages and a document of this length[18] should be acceptable to many universities. It may be used to adequately cover all the major issues which have to be addressed and in so doing show serious intent on the part of the student.

Cover page – Every university will have specific house rules as to how the cover page of a research proposal should appear. The requirements will normally include the name of the university as well as the faculty and the school. The title of the proposed research should be clearly stated. The name of the prospective degree candidate, his or her student number and his or her e-mail address are usually required. Finally the cover should include the date on which the document is submitted.

Introduction – The purpose of the introduction is to inform the reader of what is going to be researched. This requires a little background to the research objectives mentioning the problem which the answer to the research should solve. Sometimes the research question is mentioned in this section but it is also acceptable to leave the research question until the end of the literature review. An important aspect of this section is to state why this research is important and to whom it will be important. The research may be important because it might

---

[18] Although a substantial length will not be accepted as a substitute for a quality proposal, a research proposal could be rejected because it is regarded as being too short and thus not complete.

The research proposal and protocol

be able to resolve a theoretical dispute; it might produce a refinement to a theory; it might be useful to the community of practitioners; or it might influence some aspect of government policy. It has become customary to end the introduction with a list of the chapters which will appear in the final dissertation.

Academic research always has to be bounded, which means that the research has to clarify not only what will be researched but also what will not be researched. This may be referred to as delimiting the research and some researchers will write a paragraph on this into the introduction chapter.

Literature review and research question – This section of the proposal is intended to demonstrate that the prospective research degree candidate has already commenced the task of reviewing the pertinent academic literature. The academic literature is the basic underpinning structure of all academic research. This is because academic research is seen as an additive process whereby the community builds upon that which is already known. Academics are not comfortable with completely new ideas being introduced to the body of knowledge as they see knowledge expansion as being a progressive activity. In the context of a research degree proposal, this section only needs to address a relatively small amount of the literature. Those papers which it does cite should be relevant, up-to-date and should support the idea that the research question proposed is relevant to the community now. Unless it is a central issue in the proposed research, the proposal neither needs to become too involved in showing how the literature may be integrated, nor is it necessary to spend more than a nodding acquaintance at critiquing these works. This will of course be an important feature of the literature review in the dissertation.

Research design —There are many different ways of conducting research for an academic degree and these differences may be attributed to methodological choices or research design options. Most doctoral degree researchers use an empirical methodology and within that approach to research there are two alternatives which are positivism or interpretivism. A researcher may take a positivist approach and try to seek dimensions which can be measured and fit into some sort of cause-and-effect model. Hypotheses may be established and these tested. This may be referred to as a methodological choice. If this approach is taken then the researcher will be practising some aspects of deductive research. This type of research is sometimes referred to as hypothetico-deductive research. On the other hand the researcher may study an environment from the point of view of understanding what variables are present and how these variables may be reacting with each other. This would not be an hypothesis testing situation and the researcher would be engaging in a form of inductive research in order to produce a theoretical conjecture about how the individuals and other elements of the processes being studied actually function. In academic research deductive and inductive approaches are often tightly coupled. An example of this is when a deductive study tests hypotheses and rejects some of them. The results in the refinement of the theory being tested and this combination of old theory and the new insights acquired by the hypotheses rejection, may be seen as a process of induction. In this approach deduction and induction are tightly coupled.

The research proposal and protocol

An important issue within the ambit of research design is the size of the sample required. Traditionally a considerable number of informants have been sought as it was accepted that a substantial sample was required in order to give the research an appropriate level of authority. However attitudes to sample size have been changing, especially in the field of qualitative research and this has been articulated by Gummesson (1988) as follows:-

> *It no longer seems so 'obvious' that a limited number of observations cannot be used as a basis for generalization. Nor does it appear to be 'obvious' any longer that properly devised statistical studies based on large numbers of observations will lead to meaningful generalizations.*

In addition to these two major research paradigms it is possible to take other routes, one of which would combine both quantitative and qualitative approaches which we would refer to as a mixed methods approach. There are also several other less popular research designs and the prospective research degree candidate needs to state as comprehensively as possible his or her intentions as regards to how the research will be conducted.

With regards to research methods, there are numerous ways in which a researcher can undertake data collection and data analysis and these approaches are described as different research methods. Options available here range from the use of questionnaires, ethnographic studies, action research and role playing to mention only a few alternatives.

Case Study Research

In this section of the proposal it is necessary to address what will be the unit of analysis. Some case studies will look at whole organisations as the unit of analysis while others will look at individual members of staff. Other approaches to the unit of analysis could be that the researcher will focus on events such as new product launches or promotion of staff to senior executive positions, to mention only two event-type activities. The more high level the unit of analysis the more likely a qualitative approach to the research will be required.

How the data will be collected, stored, managed and eventually deleted will also be addressed in this section of the research proposal.

Executing the research design – This section of the research proposal takes a high-level look at some of the activities which will be involved in the conducting of the research design articulated in the section above. This section will consider how to find case study candidates and how to get access to appropriate informants. It is in this section where there will be some discussion of how data is converted into transcripts and entered into computers and what type of coding might be involved. There may also be some information related to the type of software products that might be used. In addition issues such as how a questionnaire may be distributed and retrieved need to be addressed here, if such a course of action is required.

In collecting data it is important to understand that the researcher will only be aware of the data to which he or she has been sensitised. This has been expressed by Ray (1993) as follows:-

The research proposal and protocol

We are beginning to realize that if we don't believe in something, it doesn't exist - no matter how much data is thrown in front of us.

Possible findings and conclusions – There is a paradox in academic research which is that if you don't know what you are looking for, you are unlikely to find it. This does not mean that the researcher needs to know in advance exactly what the findings of the research will be, but he or she needs to have some idea of what these findings will be. Also it is important that if the research concludes with something quite different to what is anticipated, the researcher will be adequately creative as to be able to understand what these findings are saying and interpret them in a useful way.

Possible limitations – Academic research is seldom if ever perfect. It is contrary to the academic ethos to consider anything perfect. Perfect is not a useful concept. There are always limitations which are due to tight budgets and a restricted amount of time or to the researcher's limited cognitive capacity; but, in addition, doctoral research is undertaken by individuals who are essentially apprentices. Doctoral research processes are learning experiences. These take individuals who have not been through the doctoral degree before and allows them to learn to be competent academic researchers. There is occasionally an individual who is working towards a second doctorate but this is fairly unusual. Those who are working towards a second doctorate or even a second masters degree are sometimes thought to believe that their first degree was in some way inadequate.

In addition, in academic research findings are always considered to be tentative, with good reason. There is always the possibility of error and this needs to be foremost in the researcher's mind. Kennedy (1979) explained this point as follows:-

Whether or not statistics are used, inferences or generalisations are always tentative, data might offer confirming or disconfirming evidence, but never conclusive evidence.

Questions to ask about the Research Proposal

- Is the research proposal good enough?
- Is the proposed research project of adequate depth and breadth?
- Is the topic of adequate interest to both the academic and practitioner communities?
- Is the project achievable in the time frame?
- Is the methodology adequate?
- Is the required data accessible?
- Is the proposed data analysis suitable?
- Does the proposed research show adequate potential for critical analysis?

Appendices – Academic research often creates a very large amount of data. This can be in the form of lengthy transcripts or it can be numerous pages of statistical tables. The main body of the dissertation should not contain these lengthy items of data. It is much more appropriate to include them, if they need to be included at all, in a series of appendices.

A research proposal has to be approved. In most cases this is by a research degree committee which may be at the Department Level or School Level or Faculty Level. In approving a research proposal the members of the committee are looking

The research proposal and protocol

for evidence that the prospective research degree candidate knows what he or she is letting him are herself in for and that they are capable of undertaking the requirements of this piece of work. Sometimes these committees can be quite difficult and ask for more explanation of some of the issues raised in the proposal. It is not uncommon for a research proposal to have to be presented two or even three times to a committee before it is finally approved. This can be a difficult time for the prospective research degree candidate as it is a process through which he or she has to go before the research can even begin.

The research proposal is sometimes treated as a hurdle over which the prospective researcher needs to jump but which is not referred to again unless the researcher wishes to make some major change to his or her research. If the topic is changed or the research question is amended in a substantial way then the research proposal may have to be revised. But this is unusual.

4.5. The research protocol

As regards the research protocol it is a comprehensive document outlining how the research will be conducted. It is more than a simple record of intent as it is an important planning tool which should become an integral part of how the research is managed. The research protocol needs to address a number of different issues concerning the elements of the research processes, practices or procedures. It is a working document and as such should be updated as part of the researcher's own time and resource management system. Working with and to a protocol gives the researcher and the supervisor some confidence that the research project is being managed and is thus not proceeding haphazardly. In addition to writing up a research protocol as described below, some researchers will also create a bar chart showing all the necessary activities and when he or she expects to perform them. These can be useful at times but the researchers should avoid the temptation to believe that he or she can live up to such a plan.

Some universities require the research proposal and the research protocol to be presented simultaneously. Other universities see the research protocol as being primarily a management document for the researcher and the supervisor. It is important to be clear as to what your university requirements are concerning the protocol. The protocol which is described in this book is a general protocol for the whole research. This will be fully explained in the following pages. However there is also an ethics protocol. This is a separate issue and the university will have separate procedures for considering and accepting a prospective research student's ethics protocol. This is not addressed specifically here. See Remenyi, Swan and Van den Assem (2011) for details of how to work with the ethics committee in order to have ethics approval granted and an ethics protocol formalised and issued.

4.6. The parts of a research protocol

A research protocol may be thought of as consisting of six separate sections. These sections are:

1. Finding the starting point - Issues related to the research question (items 1 to 6)
2. Establishing the data required - Issues related to the data and its collection (items 7 to 27)

The research proposal and protocol

3. After data collection is complete - Issues related to the management and the processing of the data (items 28 to 34)
4. What does the data mean and the theoretical conjecture - Issues related to interpreting the data (items 35 to 37)
5. What the research delivers to the community - Issues related to producing a contribution to academe and to practitioners including limitations and suggestions for future research (items 38 to 42)
6. The research ends - Issues related to presenting the work for examination and preparing for a viva and making changes (items 43 to 44)

4.7. Part 1 - Finding the starting point

The six items relating to this issue are shown in Table 4.2 below.

Table 4.2: Finding the starting point

Item No	Issue	Detail	Notes
1	Field or fields of study		
2	Research Topic		
3	Research problem or objective		
4	Literature review		
5	Research question		
6	Research sub-questions		

Before a major academic research project can begin it is necessary to establish some of the opening parameters which include being clear on the field of study, the research topic, the research problem or objective. These issues may sound self evident but it is not uncommon to find research projects which have been proceeding for some months, if not years, where these issues are not properly resolved. The literature review is the main academic route to providing the information required to resolve these issues and a literature review will also help settle on the main research question.

The research question in turn will need to be 'deconstructed' into a number of sub-questions which will allow focused questions to be asked of the informants. Sub-questions sometimes broaden the scope of the main question and this should be avoided.

In completing this section of the protocol it is advisable to refer back to the literature to ensure that focus has not been lost during the question and sub-question development.

4.8. Part 2 – Establishing and collecting the data

If the research question and sub-questions have been well articulated then establishing the data required to answer the research question should be relatively straight forward. However there are always options and the researcher will have to make choices as to which approach will suit him or her.

As mentioned in Chapter 1, case studies are increasingly popular as a method of data collection and analysis and if this is chosen then the rationale for this decision is required. Case studies require visits to the informant's organisation and this needs to be thought about as travelling is increasingly expensive, not to mention time consuming. It is not always easy to gain access to the organisations the researcher believes will be useful and thus it may be necessary to work alongside a gatekeeper who will assist with this.

The research proposal and protocol

The 21 items relating to this issue are shown in Table 4.3 below.

Even with a gatekeeper's help it still can be difficult to obtain access to those people that a researcher might wish to so it is necessary to be realistic about how much can be achieved in this respect.

Table 4.3: Establishing and collecting the data

Item No	Issue	Detail	Notes
1	Revising the literature to ensure the appropriateness of the research question		
2	Data required		
3	Qualitative or quantitative		
4	Case study rational		
5	Location/s		
6	Gatekeeper/s		
7	Data collection methods- different sources and different informants		
8	Tools required		
9	Informants		
10	Are there any sampling issues?		
11	Ethics issues		
12	Detail knowledge		
13	Arrival at premises		
14	Pre-interview		
15	Interview schedule		
16	Interviews		
17	Post-interview		
18	Other data field notes etc.		
19	Reflection immediately after the interview		
20	Expected timing of the work		
21	Cost implications		

When planning how to collect the appropriate data there are a number of factors to think about and these include:

1. deciding who should be interviewed;
2. how to gain access to the right people;
3. what sort of documents would be useful;
4. should a group interview be requested;
5. would a focus group of informants be helpful?

Many if not nearly all academic research degrees require more than one case study. When this is the case the question arises as to how to choose the different entities. Sometime researchers talk about which case studies they should have in their sample. This term sample is not quite appropriate. The case studies should be chosen with considerable care in the expectation that they will throw more light on the possible answer to the research question. One of the reasons why the word sample is perhaps not appropriate is that the few cases chosen might actually constitute the whole population relevant to the research.

It is at this point that the researcher will have a sufficiently clear view of the tasks that lie ahead thereby enabling him or her to give serious thought to the ethics implications of the research.

It is important that a researcher applies for the approval of an ethics protocol before the research commences. Some universities are very particular about this and they say that no retrospective approval will ever be given. What this would mean in practical terms is that if the research has commenced without an approved ethics protocol the researcher would be expected to go back and begin again from the very beginning.

The research proposal and protocol

The application for the approval of an ethics protocol requires one or more forms to be completed and for specimens of proposed research measurement instruments to be provided for inspection by the faculty or the school ethics committee.

At this stage the researcher needs to pay attention to the detail work required for the data collection exercise. This includes preparing interview schedules, questionnaires and beginning to make plans to look for suitable informants. Another aspect of preparing for data collection is to ensure resources are available including a recorder/video equipment etc.

The researcher needs to think about how to take field notes, how to integrate them with the other data acquired and how reflection periods can be arranged.

This part of the research protocol ends with the researcher considering the time scale of the research and also if there are any significant costs involved for which provision need to be made.

4.9. Part 3 - After data collection is complete

The seven items relating to this issue are shown in Table 4.4.

There are a number of different ways of handling data once it has been collected. Whichever way is chosen, interviews, focus groups and other spoken data collection will normally be reduced to a transcript. This is painstaking work and needs careful attention. If a technique such as content analysis or correspondence analysis is to be used then the data needs to be coded.

Case Study Research

Table 4.4: Establishing the data required

Item No	Issue	Detail	Notes
1	Creating the Transcript		
2	Tools required to analyse the data collected		
3	Coding data if required		
4	Sorting, grouping and ordering the data if required – the data analysis		
5	Version control		
6	Quantitative data – data analysis		
7	Backup arrangements		

Some researchers make the distinction between bottom up and top down coding (Symon and Cassell 2012). Bottom up coding occurs when the researcher finds the themes which are used in the coding from an examination of the transcripts created. Thus with this approach themes and their codes emerge from the researcher's understanding of the text of the transcript. Top down coding is used if the researcher has identified theoretical concepts or themes already established, usually from the literature, for the purposes of creating the codes. Some researchers will use a combination of both bottom up and top down coding. Because of the dynamic and explorative nature of the analysis of transcripts it is almost inevitable that both bottom up and top down coding will be used as the researcher's understanding grows.

If the researcher is following a program of grounded theory then coding is a central issue. Coding is an important way of labelling, grouping and summarising concepts. Sometimes when data is coded it is referred to as fractured data. On the

76

The research proposal and protocol

other hand a hermeneutic approach may be used which is more holistic[19]. Case study researchers need to be familiar with these approaches.

When using case studies the researcher needs to be aware that his or her approach to data analysis may evolve over the period of the analysis (Grbich 2007). During the analysis the researcher's view of how the data should be analysed can change. New techniques may be found and older ideas may be abandoned. Flexibility in the selection and use of analytical tools is important as is continuing to focus on how the research question may be better understood and answered.

Data is often coded on a thematic basis, although this is not the only basis on which codes may be created[20]. Codes may be used to facilitate greater understanding regarding individuals, organisations, time periods etc. One of the main purposes of coding is to allow retrieval of group data which expands our understanding.

A theme is a recurrent concept or idea which the researcher considers worth further exploration or analysis. Themes may be classified in a number of ways and they may be identified

---

[19] Holism is sometimes seen as being the opposite of reductionism. Reductionism argues that the way of solving a large problem is to define it as a series of smaller problems which may be solved separately and the sum of the solutions will represent the solution of the larger problem. Holism asserts that some problems are not amenable to reductionist approaches and these need to be studied or researched in a different way. A holistic orientated researcher will also argue that if a research problem is reduced to smaller problems than a simple summation of the findings of the smaller solutions may not adequately answer the greater or larger problem.

[20] A transcript could be coded using individual types, language groups, political affiliations and so on.

as being associated with particular individuals or parts of an organisation. In a case study there will often be many themes and one of the tasks required of the researcher is to identify relationships between and among themes. It is important that themes are distinctly different to one another. Some themes may be grouped in a hierarchical way. For example, pay negotiations, retention of staff, dispute resolution and recruitment could be grouped under a major theme referred to as human resource management. Furthermore, themes such as human resource management, manufacturing issues and sales performance could in turn be grouped as operational issues. Grouping themes into larger or broader themes employs hierarchical coding. It is important to bear in mind that themes are concepts which appear a number of times in the transcript and may be addressed directly or indirectly in the text. The identification of a theme is essentially a subjective issue related to the researcher's perception of the situation and as a result researchers will have distinctly different views about which themes exist and their relative importance.

When data is coded then there are a variety of techniques available for sorting and grouping it in order to obtain a greater understanding of what the data conveys. Care needs to be taken with the number of codes being used. Researchers sometimes employ a large number of codes and this can lead to confusion and to extra work in combining codes which represent themes which have only minor differences. The number of codes required will be a function of the research question and the richness of the data being an analysed.

Where quantitative data is involved in case study research then standard statistical techniques will be used and the results of the analysis will be incorporated into the research findings when the researcher is making his or her argument.

Data is seldom entered in one sitting and therefore different files are created representing the progress made in entering the data over time. Working with data files like this allows different versions of the data to be confused and a researcher needs to take particular care that this does not happen.

This section ends with consideration being given as to how the data can be protected against loss or accidental deletion. It is surprising how often this happens. Laptop computers can be stolen or lost relatively easily. Sometimes they are left on trains. The researcher may press the wrong button and the data is inadvertently deleted. Having proper backup arrangements if data loss occurs avoids these situations from becoming a catastrophe.

4.10. Part 4 - What does the data mean

The three items relating to this part of the research are shown in Table 4.5.

Table 4.5: What the data mean and the theoretical conjecture

Item No	Issue	Detail	Notes
1	Drawing together the different data analyses and producing an argument		
2	Proposing a theoretical conjecture		
3	Connections between the theoretical conjecture and the data		

This section of the research is regarded by many as requiring the highest degree of creativity on the part of the researcher.

If quantitative analysis has been performed then there are well established methods for declaring what the statistics suggest. This is particularly true when hypotheses have been tested. However even when more advanced statistics have been employed there is considerable agreement about what may be read from the number. On the other hand if qualitative analysis has been employed there is much less agreement and the researcher has to rely on his or her ability to craft a convincing argument based on the analysis available. When there is both qualitative and quantitative data available the argument can be strengthened if the two sources of data produce results which appeared to support each other. If however they do not support each other the researcher needs to be able to find a convincing argument as to why this should be the case.

Having reviewed the analysis and developed an argument which explains what the data means the researcher may then wish to propose a theoretical conjecture. This is the case when theory formulation is the main objective of the research. If the research has been deductive in nature and therefore some hypothesis testing has been conducted some theoretical re-formulation may be in order.

This section of the protocol describes the activities required as the research comes near the final stage of an academic degree. It is regarded by some as being perhaps one of the most important sections of the degree and therefore all researchers should be aware of these tasks from early on in the degree process.

The research proposal and protocol

4.11. Part 5 - What the research delivers to the community

The five items relating to this issue are shown in Table 4.6.

Table 4.6: What the research delivers to the community

Item No	Issue	Detail	Notes
1	Formulating the findings into management guidelines and policy criteria		
2	Confirming the usefulness of the guidelines		
3	Listing limitations of the research		
4	Providing suggestion for future research		
5	List of appendices		

Academic research is required to make a contribution to both theory and to the practice thereof. Therefore the researcher is required to develop the theoretical conjecture into a series of management guidelines which will be helpful in the everyday application of the theory. Sound academic research requires that these management guidelines be confirmed by inviting members of the business and management community to offer an opinion on them. This may be regarded as part of the validation process.

There are two other issues which need to be addressed and these are the limitations of the research project as well as suggestions for future research which have become apparent due to the research project just completed. No academic research project is ever perfect. There are always issues which could have been investigated further but which have not been, due to time or funding constraints. The researcher should comment on these at the end of the research report. Because an academic research project is always a voyage of discovery there will always be additional issues that arise which deserve attention and it is appropriate for the researcher to point these out for possible future research.

4.12. Part 6 - The research ends

Finally at this stage the researcher should prepare a list of the various appendices which will be attached to the completed work. The two items relating to this issue are shown in Table 4.7.

Table 4.7: The research ends

Item No	Issue	Detail	Notes
43	Complying with submission requirements		
44	Preparing for examination and or changes		

The final step in completing an academic research degree is to establish the university's rules concerning the submission of the document to the examination office.

This involves establishing how many copies are required and how they should be bound. It is also necessary to comply precisely with the rules regarding the text which needs to be placed on the cover of the dissertation. All of this is routine and it should be no difficulty in obtaining and complying with this information.

This chapter has described what is required to produce a research proposal and research protocol and, as can be seen,

The research proposal and protocol

these documents require a substantial amount of work. The research proposal has now become obligatory at most universities. The research protocol is required at some institutions but not others. Nonetheless the research protocol is a useful tool in that it requires the research degree candidate to think through the steps that will be required to complete a substantial piece of academic research. The research protocol should be discussed with the supervisor and guidance should be received from him or her as to how much detail to include in this document.

It is possible to conduct academic research without a proposal or a protocol. However, most if not all universities now require at least a proposal. Ethics protocols are increasingly demanded as well.

A more general protocol which is used as a planning and control document by the researcher him or herself is not yet in general use. Those who have carefully thought about the work involved in academic research and who have developed such a protocol have found it most valuable. At doctoral level academic research involves the management and control of a sizable project over a considerable amount of time. Planning makes this achievable without too many tears. If planning is not done then much of the work of a doctorate will take place in a haphazard way and if it succeeds then it will be as much a matter of luck as a matter of competence.

4.13. Summary and conclusion

Academic research is a complex activity and like all complex activities the more forethought given to it the more smoothly it will operate. This means that planning is important. However research planning should not be allowed to impact flexibility which is always important. It is difficult to foresee the twists and turns an academic research project could take as it unfolds and thus the researcher needs to be prepared to follow ideas which emerge during the research process. This is especially important in case study research because of the wide range of data sources available and the probability of the researcher being offered a number of different perspectives on the matter being researched.

Finally, some universities do not allocate a supervisor to a research degree candidate until after the research proposal and the research protocol have been written and accepted. If this is the case then a well thought through proposal will signal to prospective supervisors that the researcher is serious in his or her desire to obtain the degree. If the researcher is looking for sponsorship the same would apply to potential sponsors.

Chapter 5
Collecting the data

5.1. Data collection - a major challenge

Data collection is one of the most challenging aspects of research. It can be difficult to know what data is actually required. It can be challenging to obtain access to appropriate knowledgeable informants (as well as other sources of data) and even when this is achieved there can be problems in acquiring the data needed to understand and answer the research question. Some researchers believe that data collection is the most demanding or difficult aspect of case study research where a high degree of skill is required. This skill is inevitably learnt by doing and as such researchers should be aware that their first encounter with data collection may not produce adequate results.

The first rule of data collection is to pause before making any decisions. Too often researchers rush to collect the first type of data which comes to mind and this can be a mistake leading to a waste of time and sometimes lost opportunities. From the literature review research degree candidates will have become aware of how research questions in his or her chosen field of study and topic have been previously studied. This will point to the data which may be required for the current study. There is a strong underpinning principle in research of precedence. This means that the researcher should know how previous research was conducted and unless there is a sound reason for not doing likewise, the research should follow the approach previously taken.

5.2. Empirical research

Because a case study is empirical the researcher needs to decide who or which organisations should be approached to help supply data. The unit of analysis will also indicate which informants may be the most likely to possess the data the researcher requires.

It is a common error to think that data should be acquired from the higher echelon of the organisation being studied. Researchers sometimes want to speak to the directors due to the mistaken belief that the higher up the organisational hierarchy the better informed an organisational officer will be. Often individuals in the most humble positions will have invaluable insights to offer the researcher. Because triangulation is important in research into business and management studies it is essential for the researcher to obtain different types of data from different levels in the organisation. This may be seen as a form of informant triangulation (Stake 1995).

Different sources of data will also be important. Thus interviews are frequently the central piece of a data collection strategy but they need to be supported by organisational documents, information from the media, a focus group of users or consumers and so on. There may also be the possibility of viewing legal documents such as those available at Companies House. These different sources of data are another form of triangulation i.e. data source triangulation.

Collecting the data

5.3. Opportunistic attitude to data collection

Thinking carefully about data requirements and opportunities in advance is important but it should not be believed that it will be possible to anticipate every data opportunity which will present itself. It is essential that the researcher takes an opportunistic attitude to data collection and collects interesting items of data wherever possible[21]. However there is one note of caution in this respect which is that a researcher can become overwhelmed by a tsunami of data and this can slow down the research.

5.4. Log all data collection activities

All data collection activities should be logged in a research diary or journal. This should record when data was obtained, from whom, in what form it was received, and how it is expected to be able to assist with understanding or answering the research question. The research diary or journal is a fundamental instrument of any research project as it is the basis on which the researcher will be able to develop his or her reflections on the research project. The entries in the research diary or journal will also be the material with which the researcher can produce an audit

Data access

Academic researchers often have to compromise as to the data they are able to obtain. Sometimes it is simply impossible to find ideal informants and when this happens the researcher has to try to acquire other sources of the data which will provide the necessary insights.

---

[21] Because the universities' ethics committees require the researcher to state the sources of data which will be sought it is important to try to anticipate as many different data types in advance and have these specified in the researcher's application for an ethics protocol.

trail which helps ensure that no important steps have been omitted during the research. A research diary or journal is essentially a report to oneself of what has been achieved and this is an important aspect of the reflection process which is increasingly seen as being a central aspect of academic research.

It will be recalled from Chapter 2 that, in business and management research, case studies are often conducted within an organisation and the unit of analysis can be either the whole organisation or part thereof. The part of the organisation which could be studied may be a traditional department or division or it could be a team of people located at one site or working remotely from the organisation's offices, or perhaps even people around the world. The reader will also recall that the case studies approach may be used with several different units of analysis within one organisation or it may be appropriate to have a number of case studies conducted in separate organisations. The case study approach is flexible enough to address a large number of different or similar units of analysis. Readers will recall that the constraints on case study research have been discussed in some detail in Chapter 2.

> **Data is always required**
>
> Whatever approach to research is taken data is always required. If you are an empiricist then you need empirical data which you have to acquire from the field. If your research is theoretical then you will be looking for secondary data as well as feedback from peers through appropriate discourse.

Collecting the data

5.5. Data collection is not easy

Data collection can be difficult and it is labour-intensive and takes a substantial amount of time. It is therefore important to be as sure as possible that an organisation will fully cooperate with the case study research before you begin any on-site collection of data. However before an approach is made to an organisation the researcher should ensure that he or she has some appreciation of what the organisation does and how it operates. This will help the researcher to choose suitable case study sites and avoid researching organisations which do not have adequate potential to reveal interesting findings. There is data about most organisations in the public domain and the Internet makes some of this data available with comparatively little effort. Extra data may be obtained from the business press, trade associations, chambers of commerce and stock brokers. There is a second reason for the researcher to do some homework before approaching an organisation. Having a good idea of what an organisation does, how it operates and how it is perceived in the market place will impress the organisation's office bearers and it will encourage them to think highly of the researcher. If there is any hesitation about allowing the researcher to have access this might tip the scale in his or her favour.

Gatekeepers

Professional organisations or institutions can often be helpful in performing the function of gatekeepers. Recently a research degree candidate obtained extensive access and collected his data in record time by having a professional institute show enthusiasm for his research. The professional institution concerned introduced him to several case study organisations and helped with the arrangements for interviews with the appropriate informants.

5.6. First contact with the organisation

First contact with an organisation is of critical importance. It is quite difficult to know at what level researchers should approach an organisation. If the researcher attempts to make contact with a senior executive, he or she may be too busy to attend to such a request, which could then be ignored by a more junior person who is also too busy. On the other hand if the approach is made to a junior manager then he or she may not be able to obtain adequate permission from the organisation's senior executives for the research to proceed. As mentioned previously the use of a gatekeeper who will or may have access to different levels of management within the organisation can often be helpful and the researcher should discuss with the gatekeeper how he or she can obtain access to the individuals they believe are required to provide access to the necessary data to answer the research questions.

Because a case study is an umbrella term there are a number of different data collecting techniques available to the researcher. As mentioned before these include both quantitative and qualitative data collecting techniques. However, in general much if not most case study research is based on qualitative data and qualitative research techniques. The principal qualitative data sources and collection techniques have been previously listed in this book. Capturing data from these sources involves creating transcripts of dialogue or descriptions of situations. They include archival text as well as photographs. The range of material is substantial. Sometimes these transcripts and descriptions can be quite lengthy and they can be challenging to both create and analyse. This work will always be time consuming. If a one hour interview is recorded it can take two or three hours, or more, to transcribe this interview which will also need to be carefully edited. There is an interesting issue related to recording interviews. In general ethics procedures for academic research require the data obtained to be anonomysed and when data is collected using physical questionnaires or by having a researcher take notes during an interview, the task of anonomysations is relatively straightforward. However, if an interview is recorded on an audio device or by the use of a video camera it is really difficult to anonomyse. There are also issues involved when a questionnaire is completed online as it may not be possible to say with confidence whether the location and even the name of the informant is untraceable by computer experts.

5.7. Mixed data

One of the issues which needs to be clarified is the fact that within transcripts there can often be numbers. While interviewing an informant he or she may supply the researcher with details of some part of the organisation's activities which are best understood numerically. An example of this is a list of sales for the previous year. This could be accompanied by other numeric data such as a list of sales commissions paid to salespeople and others or perhaps numbers to do with the economy in general such as GDP or GNP. Although this is clearly numeric data it may be treated descriptively. By this it is meant that the data is not processed in any statistical form other than perhaps to be used as the basis for creating tables and graphs. The presence of these types of numbers and associated tables and graphs does not affect our categorisation of the research. It could still be qualitative research. Thus descriptive statistics are acceptable as a component of qualitative data. Inferential statistics may not be.

It has also been pointed out that the case study could have a quantitative data analysis dimension. In understanding the phenomenon of interest to the researcher, it may be appropriate to conduct a survey. For example, when data is concerning the organisation's after-sales-service, it may be valuable to establish what a substantial number of the clients feel about the service as it is offered and performed for them. Survey data of this type is often collected by the use of a questionnaire. Questionnaire design is an important issue in its own right and there are numerous text books available on this issue (See Remenyi 2011).

Such a survey would be one dimension or strand of data collected as part of a greater study strategy. Also there is no contradiction in obtaining quantitative data from inside the organisation and some quantitative data from outsiders. In general the test is whether the data can help the researcher understand and answer the research question. If this is the case then the data is useful. Researchers should not be too concerned as to whether the data could be classified as quantitative or qualitative.

5.8. Data access

As mentioned earlier it is not always easy to obtain access to appropriate individuals even when the organisation as a whole has agreed to be supportive of the academic research. There are many different types of personal reasons why a particular individual may not wish to participate in the study. One of the more important of these is that there may be some suspicion in the mind of the potential informant that the researcher is an agent for the managers or the owners of the organisation or business and that they really want to obtain information

Collecting the data

which is not normally available. Put bluntly they may perceive the researcher as a management spy. There is no easy way for the researcher to overcome this perception. Being polite and being grateful for the opportunity to speak to the individual and also behaving as naturally as possible can help build up some level of acceptance and thus trust.

But even when the informant is willing there are still challenges to obtaining useful data and these include:

- Difficulties encountered by individuals in their being able to recall events accurately. There are always problems in distinguishing real memories from vague recollections;
- Difficulties individuals have in disclosing important feelings. This is especially true when an individual describes situations which did not go well;
- Difficulties individuals have about revealing information that might reflect poorly on themselves or their superiors or their colleagues. Organisations do not like to admit that there have been problems and selective memory helps keep the reputation of individuals at bay.

It is not easy to overcome these challenges and the researcher certainly has to talk and behave in a highly empathetic way if progress is to be made in having the informant speak freely. Of course it is difficult to know if anyone is speaking freely and openly and researchers need to be aware of this.

Sometimes a second interview is required before the informant will be open with the researcher but this may be difficult due to time constraints and also the cost of a second visit to the case study site. One of the ways in which the researcher can facilitate the open conversation is by having prior knowledge about the organisation and using this knowledge to ask the informant to help the researcher complete his knowledge of how this situation actually operates.

> **Difficulty in accessing the relevant data**
>
> - Researchers need to be alert to these three areas of difficulty and produce filed notes concerning an informant's problems of recall, expressing feelings and being worried about showing him or herself in a poor light.
> - How the researcher treats data obtained under the circumstances is a function of the experience of the researcher. This is very similar to the question of 'noise'.

The researcher can also ask the informant if he or she could access organisational reports which would help illustrate the nature of the events or processes being discussed.

There are no hard and fast rules with regard to how many data types or individual informants should be included during case study research. As part of the definition of a case study is that it uses multiple sources of data, it is important to ensure that the researcher does not confine him or herself to interviews only, for example.

5.9. Sources of data

The three main sources which are available to the researcher are spoken evidence, written evidence and evidence acquired by observation.

Collecting the data

Spoken evidence - Interviewing is by far the most popular way of acquiring spoken evidence. There are a number of different interviewing formats and the researcher will sometimes employ several of these. Formal interviewing is normally the preferred method and this is a skill which many researchers do not have naturally and therefore they need to learn the skills involved. Informal interviews are also useful but it is not always easy to obtain all the information required from an informal encounter with an informant.

The second form of spoken evidence is the focus group. Focus groups, which are sometimes mistaken for a group interview, allow the researcher to obtain data both in the form of opinions or knowledge from each of the informants involved and from listening to debates and arguments between the informants present at the focus group.

By the way, there is no reason why informants need to be employees of the organisation or members of the staff of the unit of analysis being studied. Researchers are at liberty to seek spoken evidence or data from customers, suppliers or other interested parties.

Written evidence - Formal and informal reports are of considerable value to academic researchers and these should be sought wherever it is appropriate. Another important source of written evidence is communications between members of the organisation. Formal communications such as business letters and less formal communications such as e-mail or memos often contain much-needed data. But like the spoken evidence described above there is no reason for the researcher to confine him or herself to internal written evidence. Researchers may find information in newspapers, magazines, stock broker's reports, industry and trade journals.

Observation data - Watching individuals within an organisation as well as the way the organisation is structured and functions is a very fruitful source of data. It is even possible to obtain useful information by looking at the buildings, the offices, the canteens and other facilities which are available to the employees.

Often these details are missed because they are seen as being less structured than the more formal types of data obtained from interviews and from written evidence. Researchers wishing to take advantage of data obtained through observation need to be conscious of this opportunity and need to record their impressions of what they see as soon as possible. This is usually done through the process of writing field notes which can then be incorporated into transcripts later produced. Field notes are effectively memos which the researcher writes to himself or herself.

An important approach which has an element of observation data is referred to as the participant observer research. It is not entirely observation as data from conversations is also being collected. However an important part of the work of a participant observer is to perform a task within the organisation being studied and to report on it. The task is usually performed in conjunction with one or more members of the organisation being studied and the researcher records the detail of the event[22]. Obviously these people will also speak to each other so participant observer research combines observation and conversation.

5.10. Triangulation

These three sources of data may be used to support each other and thereby enrich the understanding which the researcher obtains of the organisation and the possible answers to the research question. As mentioned above the process of using data in this way is referred to as triangulation. Triangulation means that the researcher is attempting to see the situation through a number of different lenses. It has already been pointed out that researchers can look for evidence from an informant and then ask him or her to provide some sort of documentation which will throw more light on the situation. Such documentation will have two objectives, the first of these being to provide some confirmatory evidence that what has been expressed is correct and the second is to further enrich this evidence with more contextual detail. Observation can also play a role in this form of triangulation which is referred to as data triangulation.

Another frequently used lens available for triangulation is what is referred to as informant triangulation. In this case the researcher obtains evidence from a number of different in-

---

[22] Participant observer research is sometimes confused with action research. However they are quite different. Participant observer research does not require the same depth of involvement with the researched participants as does action research. Action research normally requires the researcher to lead an intervention in the unit of analysis being studied. There can also be issues related to whether the researcher as a spectator is in some sense a participant in the matter being studied and this may sometimes be the case.

Case Study Research

formants. Using informant triangulation the researcher will obtain the view of different people such as the chief executive officer, the director of production, and perhaps a chief marketing officer. These different people will have their own individual perceptions of what has taken place. There are other types of triangulation which interested readers can explore.

The mistake which is sometimes made with triangulation is the belief that if a matter is viewed from a number of different perspectives the researcher will have access to the 'truth'. Underpinning this view is a belief that evidence will converge when different sources are consulted and that any contradictions will fade as a more credible picture emerges. This is seldom correct. When there is contradictory data, triangulation does not necessarily lead the researcher to be able to establish the 'truth'. There is no simple way of resolving contradictory data. Triangulation offers a researcher an opportunity of obtaining a much richer picture of what is going on and therefore have a greater understanding of the issues involved.

It should be noted that some qualitative researchers refer to triangulation in a deprecatory way as they say that the concept is borrowed from land surveyors and as such it is too mechanistic or positivist to be used in qualitative research. This is a relatively minority view.

5.11. Reflections

The researcher's first reflections after an interview are often the most important and it is good practice to put time aside to write up the field note on each interview or other encounter with informants as soon as possible. Field notes are a key element in case study research and they should contain the re-

Collecting the data

searcher's personal impressions of what has happened during the interview. Some of the topics which need to be addressed in a comprehensive field note are:

- What is the key evidence or learning point;
- In what ways does the interview support other evidence;
- In what ways does the interview contradict other evidence;
- Have new sources of data been uncovered;
- Have new clues been offered such as new lines of enquiry, new hitherto unthought-of informants or sources of documents.

Field notes

In a sense field notes are the glue which allows the researcher to stick all the interviews together and to make sense of them as a whole. If it wasn't for the recording of impressions of the informant, his or her office and other aspects of the organisation, the interview transcript would actually be fairly isolated or unconnected to the overall research process.

It is important for the researcher to record his or her impressions of the context of the interview and what has been revealed in the interview as soon as possible.

As mentioned before, it is not possible to prescribe the number of spoken evidence collection activities that researchers should have within one case study, but it would be unusual to have less than three and there could be as many as a dozen. It is necessary to bear in mind that busy executives will seldom offer academic researchers more than one hour for an interview. In any event one hour is probably close to the maximum period of intense concentration required for a research interview. Although some researchers will find two or three interviews a day exhausting and thus hard going, others may be able to undertake four or five interviews in a day, but this rate is not sustainable over an extended period of time. It is important not to press for too many interviews.

Case study researchers need to be prepared to have a considerable volume of data to work with. It is important that the researcher be prepared both physically and emotionally for the volume of data that a case study can generate. It is not difficult for a researcher to feel overwhelmed by the data. For example, the transcript resulting from a recording of a single interview could constitute 20 to 30 pages.

5.12. A pilot case study also referred to as a field test

Before the case study research commences it is valuable for the researcher to conduct a pilot case study which is sometimes referred to as a field test. In fact in academic research the use of piloting is essential before finally accepting any particular course of action with regard to data collection. The reason for this is that it is difficult for the researcher to envisage the types of problems and issues which may occur when a particular plan is put into action or a particular measuring instrument is used. A pilot study is a formal event where the researcher attempts to find a case situation which will closely resemble the cases which he or she intends to include in the actual research project. The pilot study should be conducted with all the care and attention which will be given to the final project. The results of the pilot study should be carefully examined in order to ensure that the type of results achieved are what is required. A question which is often asked is, May the results of the pilot be included in the final data? And the answer is yes, provided the pilot study produces the type of results anticipated. In practice the approach and techniques used in the final study will differ sufficiently from the pilot study to prevent the results of the pilot study being used in the final research.

By the way, careful researchers argue that before a pilot study can be performed the researcher should conduct some pre-tests. Pre-tests are less formal than pilot tests and they tend to eliminate any early mistakes from misunderstanding.

It is a common mistake to perform either pre-tests or pilot tests and not take the results of these seriously. If these tests suggest that changes need to be made to the measuring instruments or to the procedure proposed for the research then it is quite important that this actually happens.

5.13. Cross case analysis

The issues which have been described above are relevant to case study research whether the researcher is conducting one holistic case study or multiple case studies. These points are also independent of whether there are any embedded units in the case. When a multiple case study research project is undertaken the same issues apply to each of the case studies. One of the important issues which faces a researcher undertaking multiple case study research is that the data needs to be collected from each case study site or each unit of analysis in similar ways. This means applying the same measuring instruments in each of the different cases. It also requires individuals of the same level and types of positions to provide data. In addition the different case studies need to be written up using the same format. This standardisation in approach is essential if cross case analysis is to be performed.

If this is not done carefully then there will be considerable difficulty in performing cross case analysis which is required for multiple case study research.

In general multiple case study research requires a significant degree of similarity between the different cases unless the researcher is only interested in descriptive or exploratory research in which there will be little or no cross case analysis. More will be said about this type of analysis in the following chapter.

Multiple case study research raises the question of how to select a number of different research sites. There is no simple answer to this. If an exploratory case study is involved, finding a range of different organisations in terms of industry, in terms of size, in terms of technology utilisation, this could well be a useful approach to take. On the other hand if case studies are being used in a more formal research setting then it may be necessary to ensure that there is a sufficient level of similarity among the case studies chosen. However, from an operational viewpoint, researchers often have to settle for less than ideal cases to obtain access to the organisations which they would prefer.

Researchers sometimes have difficulty in knowing when enough data is collected and for this reason they can over extend the period of data collection. There cannot be any hard and fast rules about how long it should take to collect data in this environment but it should not be extended over many months[23]. As with many of these difficult issues the answer lies with the research question. The general principle is to collect enough to answer the research question and no more. This leads to the question of when is the research question answered. The research question is answered when a convincing argument has been produced that the researcher has added something of value to the body of theoretical knowledge.

How can this be assessed? To be confident that this has been achieved the researcher needs to consult his or her supervisor and other academics of his or her acquaintance. There will be differences of opinion and the researcher needs to be sensitive to the type of answers which he or she could obtain. Recently a researcher who had collected a relatively small data set was asking PhDs of his acquaintance if he had enough data and he received, a number of times the reply, 'Perhaps'. In this particular situation it became apparent that the absence of the answer, 'Yes' probably actually meant, 'No'.

5.14. Summary and conclusion

Skill is essential for case study involving careful observation in the broadest sense of the word as well as data recording. Then the data needs to be analysed and explored in the way necessary to understand and answer the research question. The analysis of the data is not enough to do this. The results of the analysis need to be interpreted imaginatively. This will lead to

---

[23] It is possible that a research degree candidate will want to conduct what might be called a longitudinal case study but this is unusual as the time permitted for an academic degree will not normally allow for longitudinal research.

conclusions which are one of the objectives of academic research.

Sometimes it is suggested that interpretation only plays a role in research when the findings of the work have to be understood and explained. In reality there is an important element of interpretation in the task of research question formulation, data collection, data recording etc. Interpretation is central to all types of academic research and the ability to interpret imaginatively needs to cultivated. Of course imagination needs to be controlled and should not be allowed to drift into the world of fantasy.

Chapter 6
Data Analysis for a Case Study

6.1. The art of the possible

The expression 'the art of the possible' has been used to define politics. In this chapter it is used to draw readers' attention to the fact that it is not possible to deliver a comprehensive review, never mind a definitive account, of the different ways that data analysis may be undertaken in case study research. There are just so many different ways of approaching this challenge. All that it is possible to do is to provide an overview of some of the more frequently applied approaches to data analysis and discuss a number of their implications.

Before embarking on this task readers may find it useful to consider a metaphor to think about the analysis of case study research.

6.2. The jigsaw puzzle metaphor

The analysis of case study research may be thought of as being similar to solving a jigsaw puzzle. Like a jigsaw puzzle there are a large number of pieces of data or evidence which need to fit together in order to tell the story of the case study and answer the research question. The pieces of the case study puzzle will normally include interviews, focus groups, observations etc. The differences between the jigsaw and the case study puzzle is that there is usually a clear picture of the final outcome of the jigsaw on the box containing the pieces, although this is not always the case. The number of pieces in the jigsaw is also normally known in advance and there is only one way of assembling the pieces to obtain the required picture or outcome.

With case study research there is an undetermined number of pieces of data or evidence. The researcher does not have a clear target as to how the pieces should come together although it is common for him or her to have some idea as to what the final result of the research will look like.

Despite these differences the objective of both the jigsaw and the case study puzzle is to create a picture of the situation which is satisfying and convincing. And this needs to be done by fitting together a considerable number of pieces of a puzzle.

The skill of the case study researcher is to be able to bring the pieces together such that they constitute a convincing argument that the research has added something of value and there are many ways of achieving this.

This metaphor can perhaps be reassuring to a researcher especially when he or she has a large volume of data pieces of different types which need to be drawn together into a convincing picture or answer to the research question.

6.3. Answer the research question

There is no one privileged way of conducting case study analysis. There are as many different ways of approaching this challenge as there are researchers. Every researcher has to find his or her own way of analysing case study data in order to facilitate the understanding and answering of the research question. This flexibility is mostly derived from the fact that case study research relies heavily on qualitative data and thus qualitative data analysis. Addressing the nature of qualitative analysis Miles and Huberman (1994) point out that:-

> No study conforms exactly to a standard methodology; each one calls for the researcher to bend the methodology to the peculiarities of the setting.

This is not only true for qualitative research but can also apply to quantitative research as well. What is important to bear in mind is that we are always looking for patterns but not only in the data itself but also between what the data suggests and already known patterns from prior research. Much of the analysis performed in case study research relies on comparing similarities and differences between what is already known and what we are trying to understand.

A technique which is sometimes used to help in understanding the data acquired from transcripts is referred to as Template Analysis. As the name implies the researcher constructs a template of themes and codes and uses this document to facilitate the analysis of multiple transcripts from different informants. This involves analysing and presenting the data in a common format which aids the identification of similarities and differences (Symon and Cassell 2012). Used in this way a template may be seen as the building blocks of a perceptual map which the researcher aims to develop. As the research proceeds new themes can be added to the template to enrich the perceptual map. However too many themes can become problematic.

Case Study Research

When undertaking data analysis of case studies the researcher has to continually bear in mind that the purpose of the project is to increase his or her understanding so that he or she can answer the research question. The research question is the primary driver of the whole research process. Losing focus on this is a source of considerable difficulties and it will often lead to fundamental flaws in the research.

6.4. The tool box for analysis of case study data

Whatever the research question or the orientation of the researcher there are only a small number of tools available for data analysis with case study research and these are:-

- Creating a convincing narrative;
- Producing lists or schedules of constructs and concepts;
- Demonstrating the relationships between concepts using perceptual maps, flow charts, organisational charts, or tables or matrices. *diagramatically*

6.5. Creating a convincing case study narrative

One of the key attributes of a case study is the fact that its form is that of a story. This story has to be told by the researcher in an engaging manner such that it is credible to the reader. The story needs to be unfolded in a way that the lessons learned from the research are visible and convincing. If any quantitative analysis was undertaken by the case study research it has to be analysed separately and then the results of this analysis need to be integrated into the overall case study narrative. If quantitative analysis is undertaken within the case study it is used as support evidence for the main

Data Analysis for a Case Study

thrust of the research which is qualitative (Scholz and Tietje 2002).

Making sure that the story of the research is told in an appropriate way is an important part of the case study research process. It requires a considerable amount of writing skill and can account for a substantial part of the time required for the research project. Many researchers struggle with this aspect of the research and those who do not feel confident in writing at this level may need to obtain some advice as to how to improve their skills.

6.6. Producing lists of constructs and concepts

One of the objectives of case study research is to find the constructs and concepts which have a direct bearing on the answer to the research question. The transcripts produced as the output of the interviews or focus groups have to be carefully inspected for constructs and concepts. These are central to understanding the results of the research and they will be used in the perceptual maps, or tables or matrices mentioned below.

Producing lists of constructs and concepts and recording how many times they have been used during interviews with informants is the basis of Content Analysis as shown in Figure 6.1.

Content analysis is the simplest level of analysis which can be performed on a transcript or a piece of text. At this level it provides a list of issues and the number of times an issue is mentioned. Some researchers try to distinguish the importance which has been associated with these issues by different informants. This can be done by indicating how often an informant has mentioned an issue.

	Number of counts
Issue No 1	75
Issue No 2	70
Issue No 3	69
Issue No 4	47
Issue No 5	31
Issue No 6	36
Issue No 7	33
Issue No 8	29
Issue No 9	25

Figure 6.1: A typical content analysis list

If the list is further analysed to show the origin of the informants then a 2 by 2 matrix can be created which can then be used as input data for Correspondence Analysis. A 2 by 2 matrix is shown in Figure 6.2.

	No of counts	Div 1	Div 2	Div 3	Div 4	Div 5	Div 6
Issue No 1	75	50	6	3	2	4	10
Issue No 2	70	49	3	2	2	2	12
Issue No 3	69	25	10	14	10	10	0
Issue No 4	56	20	10	17	3	3	3
Issue No 5	54	30	1	0	6	8	9
Issue No 6	48	20	10	6	10	1	1
Issue No 7	33	5	2	2	2	2	20
Issue No 8	30	2	4	14	4	5	1
Issue No 9	29	0	5	12	9	2	1

Figure 6.2: A typical Correspondence Analysis list

Data Analysis for a Case Study

6.7. Demonstrating the relationships

Relationships between constructs, concepts and variables may be demonstrated in various ways including using perceptual maps, flow charts, organisational charts, or tables or matrices (Miles and Huberman 1994). Although not necessarily essential to case study research these presentations of data are very useful in explaining the research processes and the research results. In academe in general these types of physical representation of data are considered an important way of strengthening arguments.

One particularly powerful technique is Correspondence Analysis which allows a variety of perceptual maps to be created. These perceptual maps highlight a variety of different relationships which can result in a much greater understanding of the research question and the possible answers which may be proposed. Figure 6.3 shows a typical perceptual map produced by Correspondence Analysis.

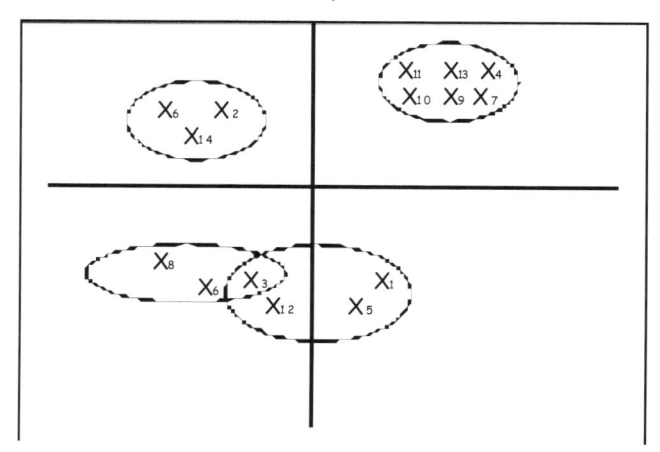

Figure 6.3: Perceptual map grouping issues with similar response profiles with a Correspondence Analysis program

6.8. Hypothesis testing

If the research question has been derived as a result of the need to test a theory then the analysis of the case study data will focus on establishing the evidence which may be used to reject the hypotheses. Much or most of the data collected in the case study will be qualitative and therefore traditional statistical applications related to hypothesis testing will normally not be applicable. Nonetheless appropriate evidence may also be found in qualitative data for the purpose of hypotheses.

Good practice dictates that both evidence which supports the rejection of the hypothesis and evidence which does not need to be highlighted. The researcher is required to weigh the resulting arguments and decide which has the greater credence. Of course this will call for a judgement and the researcher is expected to minimise his or her biases.

6.9. Theoretical conjecture development

If the research question has been derived as a result of the need to develop a theory then the analysis of the case study data takes a much broader form and the researcher seeks evidence that there are patterns in the data which could be the basis for creating a theoretical conjecture. There are multiple approaches to the analysis of qualitative data and the researcher will need to be familiar with the more important of these.

In academic research sometimes the research question is not answered as originally envisaged and the researcher may have to settle for having improved his or her understanding of the subject as well as being able to explain why the original research question was intractable.

Data Analysis for a Case Study

6.10. Data analysis begins

As already stated data analysis within the case study research framework offers a number of challenges and may include the use of both quantitative and qualitative techniques. However the first of these challenges is to clarify where data collection ends and when the data analysis or processing begins. In case study research data collection and data analysis or processing may be tightly coupled.

In dealing with quantitative data the collection and the analysis phases are quite distinct and there is seldom any problem in distinguishing one from the other. This is shown in Figure 6.4 below.

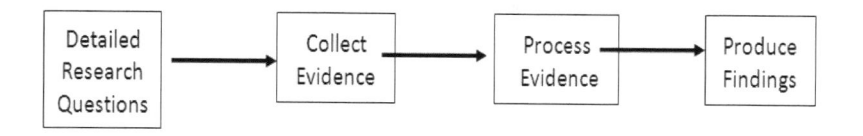

```
┌──────────┐      ┌──────────┐      ┌──────────┐      ┌──────────┐
│ Detailed │  →   │ Collect  │  →   │ Process  │  →   │ Produce  │
│ Research │      │ Evidence │      │ Evidence │      │ Findings │
│ Questions│      │          │      │          │      │          │
└──────────┘      └──────────┘      └──────────┘      └──────────┘
```

Figure 6.4: Process from research question formulation to Findings

In the case of quantitative research there is an interval between each of the activities shown in Figure 9 and the evidence/data processing only begins after all the data has been received and checked and prepared for analysis.

But when collecting qualitative data there are significant overlapping and interacting issues which allow the researcher to reflect and intervene in the research processes as shown in Figure 6.5.

```
┌──────────┐      ┌────────┐      ┌─────────────┐      ┌──────────┐
│ Detailed │      │        │      │ Reflect and │      │          │
│ Research │─────▶│ Collect│─────▶│  Improve    │─────▶│ Produce  │
│ Question │      │Evidence│      │Understanding│      │ Findings │
│          │      │        │      │     of      │      │          │
└──────────┘      └────────┘      │  Questions  │      └──────────┘
     ▲                            └─────────────┘
     │                                   │
     └───────────────────────────────────┘
```

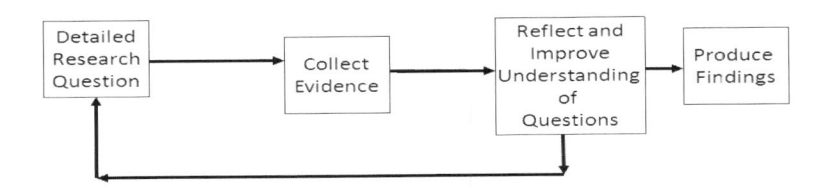

Figure 6.5: Feedback process in qualitative from research

Consider an interview. While the informant is conducting the interview he or she cannot help but be engaged in some degree of absorbing the data and coming to some conclusions as a result of being exposed to the interview conversation. In this sense the researcher's hearing the data constitutes, at least at an elementary level, a form of data analysis and understanding. In these circumstances trying to isolate data collection and data analysis is problematic. From Figure 10 it can be seen that the researcher's reflections may even influence the research questions which the researcher is in a position to amend as the research proceeds.

This point is further demonstrated by the fact that in a number of qualitative data analysis techniques the researcher is required to identify key issues for the purposes of data coding, followed by some sort of counting. Some of these issues will be suggested in the literature but others often come to mind while the researcher is collecting the data from the informants.

6.11. One case at a time

Some qualitative data researchers will argue that they prefer to collect all the possible data from the different informants and from different case study situations and then to perform their analysis. This means that the researcher has at his or her disposal the full picture of the whole data set acquired through the research and will therefore be less likely to leave something important out of the analysis.

This approach can be acceptable but there are circumstances under which it is not. The grounded theory method (GTM) suggests that if there are multiple case studies each one should be analysed in turn when the data collection process is completed. The rule is that the researcher should not commence the second case study until the first case study data has been analysed and understood. The rationale behind this is that GTM involves a learning process for the researcher and that the lessons learnt from each case study should be highlighted and internalised before the researcher moves on to the next case study. There is much sense in this approach which maximises the use of the on-going learning from the research. By the end of a substantial research project a researcher will have materially improved his or her interview technique as well as data recording and transcription skills and the ability to analyse and understand case study data.

Many researchers find themselves taking a middle position whereby they will conduct some analysis on case studies as they occur and then they will aggregate the case study transcripts and do a final overall analysis when all the data has been collected.

6.12. Finding suitable case studies

GTM practitioners point out that the researcher needs to use a technique known as theoretical sampling. The process of theoretical sampling involves looking for more examples of case study sites, if case studies is the approach being used, which will help the researcher develop a fuller understanding of the findings that are being proposed from the research at the previous research site. This means that if GTM is being used each case adds to the development of a platform on which the following case is developed. This means that theoretical sampling has at a conceptual level some of the characteristics of triangulation.

Of course case studies are used in a much broader spectrum of research than just GTM.

Case study sites may be chosen because a researcher wants to compare how a particular business and management process is operated in a large organisation as compared to a smaller one. The case study sites may be chosen to compare and contrast how processes compare and contrast in a public company with a family run business. A not-for-profit organisation could be compared with a global profit orientated organisation.

On the other hand case study sites are sometimes chosen in order to explore business and management practice in a particular industry. A number of different motor manufacturers may be chosen or different government departments selected. Focusing the research in this way may be very valuable but it may not be possible to obtain the cooperation of multiple members of an industrial sector. If the researcher can solicit the support of an industry body which will act as a gatekeeper then case study access will normally be improved.

Data Analysis for a Case Study

Whatever the makeup of the group of case study sites acquired the key issue with regards to data analysis is to insure that the organisations are in some way commensurable.

Having a small number of case studies to analyse will help with the substantial amount of work required to analyse the cases.

6.13. Writing up the case study

There is another factor to be mentioned with regard to the analysis of data acquired for a case study.

In the same way as competent interviewing implies some degree of data analysis, so does the writing up of the case study. In reporting on what the researcher has done, the data collected and other facets of the case, the researcher is performing a type of analysis. The telling of the story verbally or in writing requires data obtained in the case study site to be, to some extent, filtered by the researcher's mind. This type of filtering, which selects the important issue on which to report the research, is equivalent to data analysis.

6.14. Data overload

Case study researchers can find themselves faced with a large volume of data. It is not uncommon to have several hundred pages of transcripts as well as a number of technical reports from the organisations they have studied. These reports could be internal reviews of products or processes or perhaps marketing or human resource issues. They could also be published financial accounts, corporate brochures and other material which is suitable for analysis. Information overload can easily occur.

Researchers need to ensure that they are not overwhelmed by the volume of the data they face. Pettigrew (1990) cautioned researchers to beware of 'death by data asphyxiation'. There is usually a considerable amount of noise mixed in with the data collected during case study research and it is necessary to remove or at lease ignore irrelevant facts and figures. This is one of those skills which competent researchers learn with experience. The researcher can now create a structure for the transcripts with the noise elements removed.

Researchers need to take a systematic approach to reducing the data at their disposal through a series of steps so that they are able to distil the large quantity of data into some useful knowledge. The objective of case study research is to acquire as great an understanding as possible concerning possible answers to the research question bearing in mind that the complex research question needs to be answered within the constraints of the context in which it is found. There are many ways in which this can be done and the researcher is allowed a wide range of scope in choosing and applying his or her preferred method.

6.15. Using one holistic case study research for theory creation

From the transcript developed using the interviews together with the field notes the researchers, through the process of reading and reflecting, creates and develops a narrative. In simple terms the researcher produces a narrative which tells the story of the case. The story needs to be comprehensive which means that it should begin by introducing the object of the study and its delimitations. Then the researcher needs to describe the organisation pointing out that it is of interest as

Data Analysis for a Case Study

regards the research question, expressing why it was thought that it was appropriate for the research and then moving on to discussing the data obtained and how that data could be understood. When this story is told it may be referred to as a Primary Narrative. As mentioned above this is a distillation process and thus a transcript of 100 pages could well be reduced to perhaps 30 to 40 pages. At this stage a researcher will often produce a list of key concepts. He or she will then spend time reviewing the primary narrative to help understand whether there are any relationships between the key concepts as well as perhaps highlighting the relationships which may arise between the more important informants.

Some researchers will take a hermeneutic approach to this work. A hermeneutic approach is regarded as being holistic as it does not attempt to examine concepts on a one-at-a-time basis. This involves reading the whole narrative several times, making reference to the transcript where necessary in order to attempt to discern the meaning of the data obtained. Those researchers practising hermeneutics will read the material as a whole and record in brief their understanding of it. They will then read the document paragraph by paragraph and once again record their understanding of each paragraph. Finally the document will be read sentence by sentence. The task of the hermeneutic researcher is to take meaning out of the document at these different levels and integrate them into a comprehensive understanding of what the data may have to say. This type of research requires extensive reflection over a reasonable period of time. The researcher's reflections need to be recorded as he or she may be asked to recall the development of his or her thinking during an examination. To the uninitiated hermeneutics can appear simple, but this is not the case and this approach needs to be approached with caution. It is no mean task and hermeneutics should not be employed without skills in this respect being developed. A researcher who wishes to follow this route should work closely with his or her supervisor to ensure that a correct approach is being taken.

A researcher who does not wish to follow a hermeneutic route may make a list of the key concepts in the primary narrative. This list creation is a non-trivial task as the researcher needs to recall what he or she has heard from informants or read in the documents received or observed during data collection. The list will also be influenced by what has been discussed in the literature he or she has read. Researchers should not rush the development of this list as it benefits from a period of reflection.

With the use of this list the primary narrative is examined for the frequency with which the key concepts are mentioned. This is sometimes referred to as working with fractured data and the danger here is taking some of these concepts out of context. There is also the problem that a researcher may miss some important concepts. As mentioned before this approach is often referred to as Content Analysis, the objective of which is to list concepts and to ascertain the importance of different concepts by recording how frequently they are mentioned. Having such a list allows the researcher to further examine these key concepts and to see how they may be interrelated. The key concepts may then be converted into a mind map format which will offer more insight and understanding. An extension of Content Analysis is a technique which is referred to as Correspondence Analysis. For Correspondence Analysis it

Data Analysis for a Case Study

is necessary to create a contingency table which will list the key concepts as mentioned above as well as the source of these key concepts. By source is meant in this context the different parts of the organisation where these concepts were mentioned, or perhaps the job function of different individuals who mentioned the concepts. Correspondence Analysis is a powerful technique which allows the researcher to produce conceptual maps which not only show connections between concepts but also connections between the sources of concepts.

> **Interpretation**
>
> There is an important difference between research findings and research interpretation. The findings of an academic research project are usually presented in the form of a theoretical conjecture or perhaps a model. Then the question arises, *what does the theoretical conjecture or model mean?* In turn this question begs another which is *for whom do the findings have meaning?* It is important to be aware that different people will interpret the same findings in different ways and thus any interpretation needs to be carefully justified by the researcher.

Having produced the list of concepts which is the result of Content Analysis or produce conceptual maps which is the result of Correspondence Analysis the researcher now needs to interpret these forms of data.

In order to produce a convincing interpretation the researcher needs to create a secondary narrative. This is a further distillation of the data collected and the knowledge obtained through the primary narrative together with whatever key concept lists and conceptual maps which have been created.

Whereas a primary narrative may consist of approximately 30 pages which have been distilled from 100 or more pages of transcript, the secondary narrative should be 10 to 15 pages. What is actually happening is that the researcher is creating a dense rich picture of what the data is saying and what his or her understanding of it is. Tables and graphs are very helpful in this respect and so they need to be integrated into the secondary narrative wherever possible. The secondary narrative should contain all the important concepts which have been highlighted in the previous analysis and it should focus on the relationships between these concepts. This will effectively be a proto-theory of what the researcher has observed during the collection of data for the case study.

It has been pointed out that case studies may sometimes employ quantitative data and if this is so then the quantitative data needs to be processed and analysed in the normal way. If it is appropriate to perform hypothesis testing on the data then this should be completed and the results of these tests should be included alongside the primary narrative, the list of concepts and the perceptual maps in creating the secondary narrative.

In theory creation research the last step is for the researcher to once again distil the secondary narrative and to produce a formal theoretical conjecture. This requires intensive reflection and discourse. Many researchers find the use of graphical representations helpful here. Flow charts and/or mind maps can be quite helpful as can lists of concepts and diagrams showing the relationships between the elements embedded in the concepts or constructs (Huberman and Miles 2004).

Data Analysis for a Case Study

An example of a theoretical conjecture starts with a list of concepts or constructs such as shown in Figure 6.5.

influence	cost-justification
training	support facilities
opportunities	competitive advantage
market drivers	operations
critical success factors	approval
formulation of strategy	Environment

Figure 6.5: the principle concepts or constructs derived from the analysis of the transcript

Then Concept Flow Charts are prepared, examples of which are shown in Figures 6.6 and 6.7.

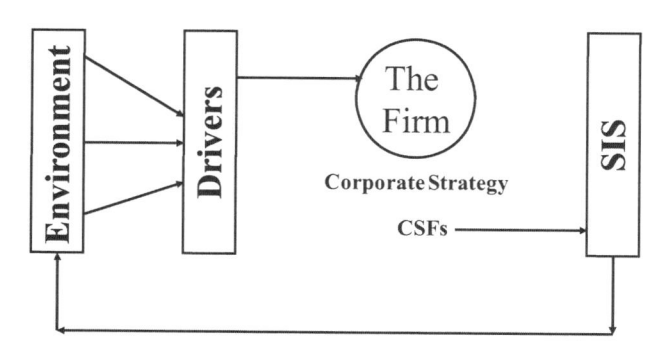

Figure 6.6: Some of the high level elements used in the Concept Flow Charts

One of the critical skills required with this type of research is that of argument building. There are a number of different ways in which an argument may be constructed but whichever is chosen it is essential that the researcher's logic is impeccable and that the argument is built on verifiable data supported by the authority from the literature.

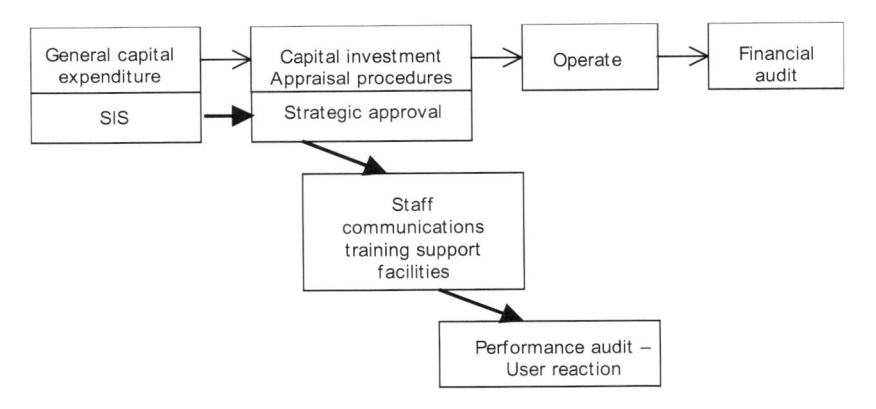

Figure 6.7: Some of the detailed elements used

The researcher should also be clear on how the theoretical conjecture produced could be tested for confirmation.

Figure 6.8 shows a theoretical conjecture which was developed for a doctoral dissertation and which arose out of the reflections and discourse based on the two flow charts above.

> *Strategic Information Systems (SIS) occur as a result of pressure or opportunities directly related to industry drivers. The firm's response to this pressure or opportunity is influenced by its strategy and by its critical success factors (CSF), and these issues determine the formulation of the SIS. The decision to attempt to take advantage of SIS is made with little attention to detail concerning cost-justification and vendor selection, but with more attention to communicating with the staff, training appropriate people and setting up support facilities.*

Figure 6.8: Completed theoretical conjecture

It should be noted that it will only be in exceptional circumstances that the researcher will produce a completely novel theory. Most researchers will be able to qualify theory in some way or perhaps enhance its scope, or modify its applicability in certain areas. It is important that the researcher and the supervisor do not have an excessively high expectation of what can be achieved during doctoral research.

6.16. One holistic case study research for hypotheses testing

When the case study is used for the purposes of hypothesis testing the researcher will have a highly focused set of questions for the informants. These questions will be worded in such a way that the researcher will be able to state that he or she was not able to find adequate support from the informants to be able to reject the hypotheses. It is critical for the researcher to be aware that hypotheses are never confirmed or proved. Hypotheses can only be rejected. If the researcher does not find adequate support for the rejection of the hypotheses then all we can say is that the hypotheses were not rejected and therefore we can accept the claim made by the hypotheses pro tem.

Case study research which is designed to test hypotheses will generally not require the type of analysis described above for the purposes of theory development.

One of the issues which arises from the use of case studies in the hypotheses testing environment is that sometimes researchers do not get clear 'yes' or 'no' answers to the questions, which is ideal for this type of research. It may transpire that informants will answer questions with the words 'sometimes' or 'on occasions' and other such generalisations. When this happens the researcher needs to interpret the meaning of these words and this does not in any way invalidate the research.

6.17. Multiple case study research for theory creation

Academic research, especially at doctoral level, normally requires multiple cases. As mentioned above an essential element of multiple case studies is that the researcher operates in a standard manner across all the cases involved. The word standard may be interpreted in a number of different ways and, as mentioned above, this can range from asking the same questions to interviewing people with similar titles. There may also be an attempt to match the organisations in terms of turnover or in terms of numbers employed or in terms of locations. Note that in multiple case study research it would be unusual to try to select the different cases on a random basis as is often the situation in survey research.

All of these different dimensions are important but perhaps the most critical is to ask the same questions in the same way. It is important that the researcher always bears in mind that the objective of the research is to get as much insight as possible concerning possible answers to the research question.

In the multiple case study situation each case study will be analysed separately and then the researcher will attempt to compare and contrast the answers from each case study site. One of the better ways of making these comparisons is to cre-

Data Analysis for a Case Study

ate a table. Table 6.1 below shows a cross case table used to compare the results of five different cases.

Table 6.1: Showing a Cross Case Analysis table comparing five different cases

Cross case analysis	Case A	Case B	Case C	Case D	Case E	
IS Strategy	5	3	2	8	8	5.2
Recruitment policy	7	0	3	8	8	5.2
Customer centricity	8	5	0	5	7	5.0
ISD Leadership	8	2	1	8	9	5.6
Staff training	9	7	3	6	8	6.6
Project Management	9	7	6	9	8	7.8
Outsourcing	7	7	7	7	7	7.0
	53	31	22	51	55	

Table 6.1 summarises the results of Content Analysis conducted on five different organisations examining seven different issues. The numeric values in the columns represent the number of occasions in which each of these concepts or issues was mentioned during the interviews with the researcher. Recall that the concepts or issues are nominated by the researcher as being the most significant he or she has heard during the interviews. Examining the table column-wise it may be seen that Case A, Case D and Case E have similar totals for the scores in the columns. This suggests that three organisations have similar concept profiles and that they are different to the other two cases i.e. Case B and Case C. Of course a simple numeric comparison such as this cannot be used to come to any definitive conclusion. But prima facie the question should be asked as to whether Case B and Case C were appropriate to include in the study.

Case Study Research

Returning to Table 6.1 it may also be examined row-wise and this leads us to suggest that the last two concepts i.e. project management and outsourcing, seem to be in general, more important than the other issues. The numbers also prompt us to ask other questions.

Another approach to comparing case study analyses is to use a verification table. This is demonstrated in Table 6.2.

Table 6.2: A verification table

Factor/construct	Evidence supporting	Evidence contradicting
Pressure or opportunities	Cases 1, 3, 4	
Market drivers	Case 1,2,3	Case 4
CSF	Case 1,2,3,4,5	
Cost justification	Case 1, 5	Cases 2,3,4

This table presents the results of a detailed examination of the data in order to verify which concepts are the most important and which concepts may have mixed or indeed no value. Reading the first row of the table it is possible to ascertain that there was supporting evidence for the concept of 'Pressure or opportunities' in Cases 1, 3 and 4. And there was more supporting evidence for the concept of 'CSF' than for any of the other concepts. It is also possible to see that on the issue of 'Market drivers' there was one case study where evidence contradicting the support for this concept was found and when it came to the concept of 'Cost justification' there is the possibility of more evidence contradicting the importance of this concept. Of course a simple counting of the number of cases which support the concept as opposed to contradicting a concept is not adequate as a basis for coming to any conclusion. Reading this table it is interesting to note that with regard to the first concept 'Pressure or opportunities' it appears that there was no mention of this in Case 2 or in Case 5. This should provoke the researcher into wondering why this was so.

The analysis of data or evidence is always a blend of creativity and analytical skill and the cross case analysis table and the verification table are simply tools which can trigger ideas in the mind of the researcher. By the way there are numerous opportunities for researchers to create tables such as the two discussed above and researchers should experiment with many different ways of presenting the data they now have at their disposal. As illustrated above in this chapter tables, figures and graphs are central to the understanding of complex data (Huberman and Miles 2004).

The credibility of the argument is a function of the researcher's skills as an academic writer (Stake 1995). It is always critical that researchers write well but this skill takes on an increased importance in case study research because one of the central features of the case study is the story or narrative which delivers the knowledge developed from this type of research. It is important to take the time to write case study research in an engaging and convincing style.

There are a number of important research issues which are regularly mentioned in conjunction with case study research. These are validity, reliability and generalisability.

When conducting a study research, validity will become apparent from the way in which the research question and sub-

Case Study Research

questions are phrased. It is important that these are field tested to ensure that they will be appropriately understood by the informants. An error which commonly occurs is that the sub questions do not constitute a means of probing the issues embedded in the main research question, but are different concepts which may extend the scope of the original question. This should not be allowed to happen. Another issue which may impact the validity of the research occurs when the concepts or constructs used to obtain data are not unambiguously defined. Any confusion between the researcher and the informants will have a direct impact on the validity of the research. Other issues of validity will become clear in the narratives which are written as the transcript is worked to become the primary narrative and then the secondary narrative. Researchers would have to be careless to produce research which does not have an adequate degree of validity.

With regard to reliability, much of the research conducted in social science has difficulties with this issue. Except when working in the control environment available in the laboratory, it is difficult to replicate precisely the circumstances of the research. Conducting a case study for a second time does not provide the necessary control. Even if it were possible to revisit the case study site and speak to the original informants, their views would no doubt have been coloured by the first experience. If the researcher was to approach a different set of informants, once again expectation can only be of obtaining a different result. Of course this does not even take into account the fact that the researcher will be different from having been through a first pass of the research. Despite this there is an expectation that if the research is conducted in a similar organisation within a similar context the findings will

Data Analysis for a Case Study

not be completely different. This is probably as far as one can go with regard to the concept of reliability in case study research.

Generalisability is an issue which has troubled case study researchers. An important tenet of research is that something of importance has been added to the body of theoretical knowledge. In order to assert its importance it is usually considered that the new element of knowledge will be applicable to a range of different organisations in different situations. In social science in general considerable care has to be taken when thinking about the issue of generalisability. The first point to be made in this respect is generalisability is not a binary concept with only two possible conditions i.e. the findings are generalisable or the findings are not generalisable. There are degrees of generalisability. So it is possible to say that a particular result will have relevance and applicability to some degree in a number of organisations. It is of course advantageous to specify the types of organisations and the extent to which the researcher believes that there will be relevance and applicability. The second point which needs to be remembered is that it is as ridiculous to claim that a particular finding is universally generalisable as it is to say that it has no generalisability whatsoever. Some researchers would argue that the word generalisability has no place in case study research. They would assert that the concepts of authenticity and transferability are the words which should be used and which of course convey a not entirely different meaning.

From the above it is clear that there is a considerable amount of subjectivity in case study data analysis. In research subjectivity is associated with bias which is regarded as an inhibitor of rigorous and reliable research findings. For this reason some researchers are very cautious about the case study approach to research. But all researchers have biases as Gould (1988) reflected:

Science is not an objective, truth-directed machine, but a quintessentially human activity, affected by passions, hopes, and cultural biases. Cultural traditions of thought strongly influence scientific theories.

In general case studies do not present significantly more bias than other approaches to social science research but nonetheless researchers need to be conscious of this issue and be on their guard to avoid bias wherever possible.

Data analysis for a case study is as much an art as it is a science. There is no privilege route or via regio to interpretation or understanding.

It calls for attention to detail but it also calls for the creation of a convincing argument that the researcher has been able to add something of value to the body of theoretical knowledge. This type of data analysis is a skill which does not come naturally to individuals and it is important that researchers spend the time to acquire and hone the skills before they can consider themselves to be competent case study researchers.

6.18. Summary and conclusions

This chapter has been wide ranging as it attempts to give a high level overview of the many different issues involved in data analysis for a case study. Only a relatively small number of options have been examined in a little detail. There are many tomes published on this subject and researchers need to familiarise themselves with the considerable detail required to be competent in that field.

As should be expected data analysis is a major part of the work in using case studies as a research strategy. Qualitative data will generally be reduced to a transcript and analysis using some form of qualitative data analysis. The actual approach will depend on the preferences for the researcher who will normally choose between a hermeneutic style or a content analysis type approach. The hermeneutic style approach will attempt to provide a more interpretivist understanding by addressing the text directly while a content analysis type approach will look for patterns in the words used and in so doing attempt an understanding based on some form of number analysis.

If quantitative data is used in the sense of inferential statistics then this will be analysed in the normal way. When qualitative and quantitative data are used in the same study the results of their different analyses need to be integrated in the write up of the argument which presents the findings and the results.

Case Study Research

Chapter 7
The case study writer as a story teller - some style and form issues

7.1. Introduction

This chapter is about some of the issues related to the story telling or narrative aspect of a case study write up. It is not its intention to address the detail of how the case study should be presented as this is addressed in Chapter 8. Writing up the case is so important that it is necessary to address it through two different chapters which look at style, form and layout versus content issues.

After all the data has been collected, analysed, reflected upon and the researcher has reached some conclusions the case study has to be written.

Some researchers say that the most difficult part of case study research is the writing up because this requires the ability to focus the analysis and the synthesis of the data, interpret it to present a conclusion and then reduce this to a narrative which engages the reader.

With these in mind, the work of writing a case study is indeed substantial if not a daunting task.

7.2. The importance of the write up

It is important to note that the case study is more than a simple description or account of the field work. An academic case study normally requires the researcher to produce a report of the research from the problem definition to the interpretation and then to provide recommendations to the research community. Some researchers report that they cannot claim to have a full understanding of the material which they are studying or researching until they have committed it to writing and thus clarifying for themselves what they have done and what they think about it. It seems that the cognitive processes involved in writing can directly aid and even enhance the writer's understanding.

7.3. The transcripts are not the case study

During the data collection phase of case study research data will have been gathered from several different sources. These often include interviews, field notes, focus groups and descriptions obtained from observation. These transcripts are not the same as the case study write up. Writing up the case study will draw on these data and sometimes quote verbatim from them. But the written case study is a much fuller account of the research which draws on the total experience of the researcher's involvement with the case study material and thus it is much more than simple extracts or examples from data collected.

7.4. Writing as a craft skill

At the outset it is important to realise that successful writing and especially academic writing is quintessentially a craft skill. This means that it cannot be learned quickly, nor can it be learned out of a book or by attending a seminar or workshop. Good writing is learned slowly by the novice working alongside an accomplished practitioner. Some universities have begun to hold writing retreats where those learning to write will be ensconced with experienced writers for a number of days' intensive instruction and practice. Unfortunately this option is by no means available to everyone.

More typically where research degree candidates are involved the supervisor will offer his or her skill and knowledge to the process of preparing drafts of the written material. The number of drafts required will vary, but novices should not be surprised if their writing needs to be reworked several times. If a novice academic is preparing a paper for a journal, having a previously well published co-author who actively assists with the writing will be to some extent equivalent to having an accomplished practitioner with whom to work.

It can take some years to develop the writing skill required to be well regarded as an academic writer and novices should not be discouraged if they do not have instant success. It will only be exceptional individuals who can immediately adopt the writing skills required.

7.5. The narrative nature of the case study

The first step in preparing for this work is to focus on the narrative nature of the case study. By narrative is meant that the case study has to read or be told as a story. Stories are vehicles for information and understanding to be transmitted throughout society and a story allows a holistic account of the research to be provided. Academic researchers are not formally accustomed to telling or writing stories as part of their research work. This is despite the fact that story telling is an intrinsic part of the way they relate to one another on a daily basis. But of course, the stories told about our day-to-day experiences do not have to stand up to close critical review. They do not have to be expressed formally with considerable care being taken about the precision of the language used. Slang may be used in every day stories and some people use what is referred to as expletives, which are totally unacceptable in academe.

Academic narrative or story telling has to be accurate, focused on the research question, precise in detail while being parsimonious, free of slang and expletives, devoid of any ambiguities or misleading statement, and lead to an argument that the research has successfully contributed to the body of theoretical knowledge. The written case study has also to demonstrate the scholarly achievements of the researcher.

These narratives are intrinsically complex and for this reason this type of work takes time to conceptualise and then to transfer the ideas to paper. It is seldom possible to rush good writing.

7.6. Styles of the narrative

There are several orientations which the narrative may take of which the most frequently used is that the researcher is an objective reporter and interpreter of the research processes conducted for the case study. In this approach the writer uses the passive voice and does not appear to be directly expressing a personal opinion. Here, case study is the accumulation of objectively verifiable data and the objectively reported interpretations and opinions of the informants.

The case study may also be developed in what can be regarded as an account of the researcher's personal experiences with the organisation or the group or the event which is the focus of the case study. Here the active voice of the researcher is used to recount what happened and what was learned by the researcher. However, although more personal in its presentation, objectivity and accuracy is still claimed.

Some case study researchers suggest that as it is nearly impossible to sustain a detached and objective point of view and that the data supplied by the informants will also be influenced by their knowledge of their situation being subjected to a research program, the pretence of objectivity should be abandoned. In general this view is not widely held, but it is important to note that some researchers take this position. A case study written from this point of view will be far more orientated to subjective impressions of the researcher and the informants.

7.7. Important dimension of a case study write up

Although there is no one single road map with which to guide the writing of a research case study there are some important principles that should be followed, or dimensions which need to be present if the text of the case study is going to have adequate credibility and impact. In this respect the writing of the case study needs to:

1. be bounded in terms of time, location, environment, issues including the research question, people and outcome;
2. be contextualised with a starting date, and the place of the story needs to be established;

3. introduce the principle characters with care and avoid mentioning too many characters simultaneously;
4. explain the environment or the context of the case;
5. address the main issues in sufficient detail that the reader is clear about the main purpose of the narrative; when the main issues have been unfolded then the reaction of the principle characters and other players needs to be described. Points 3, 4 above could be regarded as the plot and this point 5 could be regarded as the outcome;
6. reflect on this outcome and how it could be interpreted;
7. map the outcome back onto the research question;
8. ensure that the narrative describes the resulting learning points;
9. suggest how the findings of the case study can be used to good effect in practice.

Within the above nine points the researcher has much freedom of expression and therefore the advice given in that respect, needs to be of a general nature. Perhaps the most important single piece of advice is that *the writer has to remember at all times, the audience for which he or she is writing is a community of academics and that this audience is primarily interested in understanding how the research question is being answered*. Any other information supplied may well be superfluous.

7.8. Written in an engaging manner

The text needs to be written in an engaging manner. This usually means that the writer needs to search for an opening sentence and a way of unfolding the story which keeps the reader's attention. In doing this it is important that the researcher writes in his or her own voice. The words used should be understandable by most academics and the metaphors used should be carefully controlled so as the language does not appear affected or pretentious.

The opening of the work needs to be followed by explaining quite quickly, without being too abrupt, what are the main issues in the research. This comes naturally to some writers while others, as mentioned above, have to struggle with this and may have to redraft their work many times. Sometimes advice is given that sentences should be short, but it is possible that sentences can be too short giving the impression that the work is written too trivially. It has been said that changing the lengths of sentences and paragraphs can keep the reader's attention. However, this is too mechanistic for the quality of writing required for academic research, where vocabulary and turn of phrase is more important, but obscure words and phrases should not be used. Although there is little or no room for imagination in dealing with the data, the writing up can be considerably enhanced by creative and imaginative writing. Some novice case study writers will spend a considerable amount of time trying out different word and phrase combinations with regards how to start the case study and thus capture the reader's attention.

7.9. Care with language

The academic writer is required to minimise, if not entirely eliminate, the use of words and phrases for which there are no fully agreed definitions. Words such as dearth, plethora, fantastic, unbelievable, etc. should not be used. Phrases like 'it is clear' or 'it is obvious' or 'everyone knows' or 'boils down' or 'middle-of-the-road' are not acceptable. These are but a small example of the many words and phrases which need to be avoided. This type of language is more akin to journalese, than academic communication. Avoiding these types of words is part of the rhetorical convention under which academic researchers need to work.

Many writers have an almost subconscious tendency to include unnecessary adjectives and adverbs in their writing. The use of the word 'very' is often superfluous to the story being related or the argument being crafted and generally should not be used. Before finalising the case study write up the researcher needs to check each sentence to ensure that there are no unnecessary adjectives and adverbs. Checking for adverbs is relatively easy as the researcher can use the search function for the suffix 'ly' and this will highlight most of these unwanted words.

7.10. Different story frameworks or forms

van der Blonk (2002) pointed out that there were at least four distinctly different story frameworks or forms available to use for case study writing. These frameworks describe how the story could be presented to the reader of the research. It should not be thought that the choice of framework is simply a matter of the personal preference of the research. The choice will be influenced by the research question and the researcher's basic approach to the collection and understanding of the data acquired. It is possible to combine these forms when the need arises.

The four frameworks described are:-

The case study writer as a story teller - some style and form issues

1. a chronology or timeline approach;
2. a play format;
3. a biography;
4. a recollection of voices.

These are not the only forms of storytelling available, but they are the most frequently encountered.

7.11. A chronology or timeline approach

Many if not most case studies have a strong chronology or time line underpinning to them.

When using the chronology or timeline approach the narrative is carefully unfolded in terms of a series of time frames which means that the story is unfolded in terms of events which happen across a continuum of time. The events of each time period be it a day, week, month or year, are described in terms of what happened, who was involved and what the outcome of the event was. In such a case study it would be important for the time dimension to be made clear on a continuous basis and that all the contextual information should be described in terms of the dates when they occurred or became important to the narrative. Analytical chronologies can be important in considering cause and effect relationships. When identified in this manner it may become easier to recognise cause and effect. Of course the researcher has to be on the lookout for specious relationships which are incorrectly claimed to be cause and effect due to the *post hoc ergo praetor hoc* argument.

7.12. A play format

A case study written using a play format will have different and distinct sections, which will correspond to an act in a play. It will be necessary to set the scene or scenes for the characters of the case study and to allow their voices and the interaction of their voices to be heard. Then there will be the reporting of the data obtained from the principle informants through their individual voices. This will be presented as summaries of the transcripts of the interviews and the other spoken evidence. Effectively the play format incorporates both story telling descriptions in the setting of the scenes as well as the actual voices of the informants.

The strength of this approach is that it allows scope for detail to be aired as a discussion between different informants, such as that obtained in focus groups. It also provides for the opportunity to introduce, in a relatively elegant way, other comments from subsidiary sources to the main arguments.

7.13. A biography

This form of case study is useful when the research focuses on how individuals played major roles in the events which have been studied. All the necessary context has to be provided, but the focus is on the individuals involved. Biographical case studies may be used in the context of longitudinal research with the individuals' roles, motives and consequences being included as a central part of the case study. Biographies are by their nature highly subjective in that it is not possible to recount all the aspects of anyone life story and thus what is included and not included in this type of narrative determines its impact. In addition people behaviour is always subject to misrepresentation and an academic biographer needs to be

The case study writer as a story teller - some style and form issues

careful that any account provided in a biography is well balanced. This form of case study is challenging and only experienced researchers should consider this option.

7.14. A recollection of voices

Some researchers believe that the voice (i.e. the words) of the informants is by far the most important source of data that can be acquired during a case study. Furthermore, it is suggested that if the voices of the informants are simply reported then there is much less likelihood of bias creeping into the research. If this is so then basing the write up of the case study on the voices of the informants is the most authentic way of delivering the data and perhaps even the knowledge acquired from the research. If this approach is taken then the voices generate the story of the case study.

A variation of the voices approach is where the researcher describes the case study in his or her own voice and then draws on extracts from the voices of informants to support his or her descriptions, claims and arguments.

Here the write up would include a material number of direct extracts from the transcripts acquired during the research process. Of course other contextual data would also be supplied and this part of the case study would be recognised as being the "voice" of the researcher.

7.15. Not mutually exclusive

The above four frameworks are not mutually exclusive. A case study could for example be produced primarily as a play, while incorporating some aspects of a biography and a chronology.

Any combination of these four forms may be used and they may be combined in any order.

7.16. Summary and conclusion

It has sometimes been mistakenly claimed that case study research is an easy option. This is far from being correct. For the reasons described in this book case study research is complex. Furthermore it requires a high level of dexterity with language in order to produce a convincing argument that the researcher has added something of value to the body of knowledge.

On the other hand the challenges are not that great and some researchers report that they have found case study research to be particularly satisfying and rewarding.

The case study writer as a story teller - some style and form issues

Chapter 8
The case study dissertation

8.1. Case studies for research degrees

Case study research may be used to conduct research for the purposes of publishing a paper in a peer reviewed journal. However most case study research is used in the context of a research degree, normally masters or doctoral degree[24]. A case study could be presented as the research design for a masters degree. For a doctorate it would be more common to use a number of case studies, say three or four, although doctoral degrees have been awarded where only one case study was used.

This chapter addresses the issues which need to be considered when using a case study design for the purposes of a research degree.

8.2. Preparing the ground for the case study write up

The form of a case study dissertation or thesis may not differ substantially from other dissertations using a different research design but there will be some differences which need to be understood and kept in mind.

A dissertation or thesis should always tell the story of the research. This story ranges from statements and arguments as

---

[24] Case study research can also be used for undergraduate research and is becoming more popular at this academic level.

The case study dissertation

to why the issues being studied are worthy of academic and practitioner interest, as to how the research was performed and to the conclusions arrived at. This story needs to be carefully told or unfolded in such a way that the researcher convinces his or her examiners that this work has resulted in something being added to the body of theoretical knowledge. When a case study research design is used there is the story of the case within the story of the research. At a doctoral level normally there will be multiple cases used and thus there will be a number of stories from the case study sites which will be combined and utilised within the overall story of the research.

There is some choice as to where the case study write up is placed within the pages of a research degree dissertation. If only one or two cases have been used then it may be possible to locate the case studies as separate chapters in the body of the dissertation. If more cases have been researched[25] then it is usual to include synopses of the case studies in the appendices of the dissertation. A researcher should consult the rules of his or her faculty or school to establish what the house rule is. It is also good practice to look at previously accepted dissertations in the faculty or school library. Be sure to establish the house rules of the university or business school where the work is to be submitted.

Case study researchers will benefit from consulting a specialist text on the subject of writing up their research (Remenyi and Bannister 2011).

---

[25] The number of case studies undertaken for a doctoral degree should be limited to perhaps four or five.

8.3. Chapter One – The Introduction

The first step in writing up any dissertation, including case study research, is to be clear on the issues being studied and why these are worthy of academic interest. Of course academic interest is not on its own enough as there is also the need to show that the researcher will deliver practical insights which may allow organisations to perform more efficiently or effectively or both. This first chapter should also make clear the context in which the research is being conducted and explain any special issues which will be involved.

A list of the remaining chapters with a two or three sentence summary of what they will contain is often provided in Chapter One.

It is important that this chapter makes clear the researcher's enthusiasm for the research topic.

8.4. Chapter Two – The Literature Review

The literature review will follow the typical approach and will need to be as thorough as usual. As part of a case study based dissertation it would be expected that the researcher found other published studies which have also used case study approaches and these would be critically reviewed in this chapter.

It is important that this chapter is more than a simple historical account of what others have thought. The researcher needs to demonstrate his or her ability to be able to make sound assessments of the value of former thought on the topic being researched.

The case study dissertation

The main research question and sub-question need to be stated in this chapter and these should highlight concepts or variables or constructs required for the research questions.

8.5. Chapter Three – The Research Design

Some researchers will argue that there is an unspoken assumptions in academic research that it is better to follow in the footsteps of previously successful research designs rather than pioneer new design strategies. If a new research design is to be adopted then a justification for this should be provided.

This chapter will discuss in some detail the advantages of using the case study approach. Other approaches should also be addressed and it should be made clear why these different approaches were not considered better or even as good as the case study research design. This methodology chapter is also used by the researcher to show his or her philosophical stance or orientation and how this links to the research design chosen.

The research protocol will be described here as will the work which was required in obtaining approval from the ethics committee.

8.6. Chapter Four – Executing the Research Design

It is in this chapter that the researcher describes the work conducted in operationalising the case study research design. The number of case study locations or sites addressed will be stated with a high level synopsis of the work produced.

Some researchers will include summaries of the interview transcripts in this chapter, although it is more common to

Case Study Research

supply these, especially if they are of length, in an appendix to the dissertation. Full transcripts are often too long to include in the dissertation itself so high level summaries are normally provided. It is important to make sure that the names of the organisations and the individuals who have participated in the research be anonymised in any transcript or summary thereof published in the dissertation[26].

Here the researcher's skill at creating high level narratives which describe the cases studied is most important. Chapter 7 outlines the main options which are available to the researcher in this respect.

Also it may be recalled from Chapter 6 that flow charts, perceptual maps and tables are important in demonstrating how the researcher has moved from the data, i.e. the transcripts, through the narratives to the theoretical conjecture. This is the heart of the research. Specific guidelines cannot be given as the way this process unfolds will depend upon the research question and the various different data sets which the researcher has obtained.

Sometimes data will be coded and counted and sorted and then recoded into higher level codes. The higher level codes

---

[26] According to the Information Commissioner Office in the UK "Anonymisation is the process of turning data into a form which does not identify individuals and where identification is not likely to take place". University ethics committee have generally adopted the stance that research data should be anonymised as soon as possible and only used in this form. However some organisations are actually pleased to give their consent to have their name used in case study research in a responsible way and where this is obtained there should be no problem with using the name of the organisation.

The case study dissertation

will in turn be counted and sorted as the researchers move towards a theoretical understanding. This is in effect a form of Content Analysis which is regarded by some researchers as too mechanistic.

A different approach to case study data is to use a hermeneutic style analysis. Hermeneutics involves reading and rereading the case study transcripts and other collected data and through the process of reflection arriving at an understanding of the 'message' contained in the text. Hermeneutics requires the texts available to be read holistically and then to be read again section by section and even paragraph by paragraph to seek out different ways of understanding the situation being studied.

In general it could be argued that hermeneutics requires a higher level of reflective types skills than the content analysis type techniques. It is more suitable for experienced researchers than novices.

Researchers should keep a detailed audit trail which may be logged in a research diary of how they made the transition from data to theory. There may be a number of steps involved in this process and the researcher should be aware that he or she is permitted to change his or her mind as he or she compiles this research argument from the numerous sources of data that are available. The jigsaw metaphor is the most appropriate here and sometimes the remaining pieces of the puzzle may require the research to 'break up' the understanding which he or she thought they had and rejoin the pieces in a different way. The jigsaw puzzle in not finished until the last piece is in place.

Case Study Research

The audit trail should be reported in this chapter. This is an intellectual audit trail which describes the thinking of the researcher as he or she moves from one research activity to another. It is not necessary for the activities to be described although it is sometimes useful to so do. What is required is to report how the researcher's mind set changed over the period of the research.

This is a particularly important part of case study research and researchers should be aware that the development of this chapter takes time. It will seldom be less than a couple of weeks' work and it may be some months' work. Researchers often report that this is the chapter which gives them most satisfaction when it is successfully completed.

8.7. Chapter Five – Conclusions of the Research

The researcher needs to argue that he or she has added something of value to the body of theoretical knowledge and it is in this chapter that the argument needs to be crafted. The first step in this process is to state the findings of the research and then to point out that the findings have made a contribution to our understanding of the subject being researched. It is then usual to express this contribution as a model or perhaps a theoretical conjecture or both. Perceptual maps, flow charts, organisational charts, or tables or matrices can be useful here.

Other issues which may be addressed in this chapter include:-

- Deciding how to illustrate and support the findings by numeric, diagrammatic or narrative data;
- How are the findings linked if at all?

The case study dissertation

- State the different ways of viewing these issues i.e. different perspectives;
- What perspective is preferred and why?
- What evidence is there to support this preference?
- Are there any potential problems with this perspective?

This is a particularly important chapter and it requires time for reflective thought.

8.8. Chapter Six – Reflections, Limitations and Suggestion for Future Research

Increasingly researchers are required to show that they are capable of some degree of reflection concerning what they have been doing and how they have conducted their research. No research project is ever perfect as it is intrinsically a learning experience and therefore how the researcher would approach this work again with what has been learnt is always revealing.

During his or her voyage of discovery the researcher will have found other issues which could be given attention and it is customary for a research dissertation to include such suggestions (Remenyi and Bannister 2011).

8.9. Appendices to the dissertation

In case study research the appendices are used to provide versions of the narratives which have been developed from the transcripts. Of course other relevant material will also be included in this section.

8.10. Summary and Conclusions

In order to write well a researcher needs to be in command of the research area, the research processes and the contribution which the research has made, and this is especially true with case study research. If there is any question of the researcher not having completed the research to the required standard this will become apparent in the writing up of the work. It is important that the writer does not over-sell the achievements of the research but rather presents them in a positive but balanced way.

There are several different styles of writing up a case study. None of these is regarded as intrinsically better than any other. The researcher needs to find the style with which he or she is most comfortable and to develop that approach. Sometimes the styles are combined into a hybrid form.

As may be seen above, a case study based dissertation has much in common with other academic dissertations but there are also some important differences. It is important that these are recognized, understood and taken into account with the writing up of the research.

The writing up of the research is challenging and should not be left to the very end. With the technology available, writing up research should be an on-going issue which can be started as soon as the research begins. This has the great advantage of preventing a rush job when the research draws to a close.

Chapter 9
Pilot Studies or Field Tests for Case Study Research

9.1. Introduction

Pilot studies or field tests have already been mention briefly in Chapter 5 and this chapter provided more detail about this important issue. Some researchers have not given this aspect of their research adequate attention and as a consequence have found that fatal flaws have crept into their work or have been present as misconceptions from the beginning. Pilot studies are important and need the full attention of the researcher.

A pilot study or a field test[27] is a small-scale implementation of a larger project which is under consideration. The purpose of a pilot study or field test is to identify potential problems which could arise and make arrangements for these to be managed. A pilot study is an investigation in which the challenges offered by the research process are considered and the proposed approach is improved so that the research can be conducted successfully. In business and engineering circles this is also sometimes referred to as a proof of concept project although a pilot study would often deal with more detail than a proof of concept report.

---

[27] Increasingly the term field test is being used in place of pilot study to mean the same thing.

In academic research a pilot study usually focuses on the research methodology including the data required, the possible informants, data collection, data recording and management as well as data analysis activities. However other issues in academic research may also be considered in a pilot study.

9.2. Refining the research design

A pilot study should result in a researcher being able to refine his or her research design and for this reason the researcher needs to listen carefully to the feedback obtained during the pilot study and to implement relevant suggestions that represent improvements. Pilot studies are sometimes seen as just another tedious bureaucratic obstacle to beginning the real research and researchers can rush this aspect of the process. This is a mistake as a pilot study can identify problems, which if not attended to, could make the whole research project invalid.

In academic research a pilot study focuses on how the research methodology will work in practice. There can be a substantial gap between what is considered possible during the research methodology conceptualisation phase and what can be achieved in the field. It is the function of a pilot study to find and highlight any such gaps. The pilot study may also address other issues in academic research that need careful consideration such as the realistic choice of research informants, the time required for interviewing and travelling and the work required in transcribing any recordings made.

9.3. Data requirements

In the context of academic research, once the research question and sub-questions have been settled the next step is to consider what data is required to answer these questions. In the case study method there will be multiple types of data that will contribute to this objective so the researcher has to decide which data types are most accessible. Some data sources will be internal to the organisation or group studied while others will be external and maybe even subject to public record such as financial accounts can be.

Sometimes this aspect of the pilot study is readily answerable. For example, no matter how desirable it may be in answering the research question, it is not likely that an academic researcher will obtain access to the personal diary of an Executive Chairman of a large bank. On the other hand an academic researcher might, with the right network, obtain a schedule of the meetings held for a week by an Executive Chairman of a large bank. The minutes of senior management meetings might be of considerable interest, but are seldom available. On a different level, although it may not be possible to obtain a useful sample from managers employed throughout a particular industry, it might be possible to obtain such data from one or two appropriate organisations and this needs to be known before the data collection process for the research formally begins. A pilot study would address such issues.

9.4. Data collection

The first activity in the pilot study is to collect some data. This is effectively attempting to test the design of the measuring instrument/s and also, especially in the case of interviewing, to test the techniques of the researcher. It is important that these activities be done in an environment, and with informants, similar to that which the researcher will encounter when the full-scale research is undertaken. It is common for the researcher to have to re-word questions in either an interview schedule or a questionnaire as a result of this exercise. The researcher may find that his or her intended approach to the interview has not worked well and needs to be rethought.

The next step in a pilot study is to examine how the data will be managed, processed, analysed and interpreted.

9.5. Preparing data for analysis and interpretation

Assuming that what appears to be appropriate data has been acquired then the question arises as to how the data could be prepared for analysis. This will usually involve the use of interview notes, field notes or memos, recordings – either audio or video and possibly questionnaires. Any of these collection techniques will produce a large amount of data and care has to be taken that it is stored appropriately. A data filing and retrieval plan is needed for this and the pilot study should address this. Data will normally have to be entered into a computer, which can be a substantial task, and how the researcher will do this requires planning. Often data coding will be required and the pilot study should consider how this will be done. A pilot study should examine the suitability of the proposed data for the analysis that will hopefully lead to understanding and answering the research question and sub-questions. This can be a significant part of the pilot study.

9.6. Checklist

For a comprehensive pilot study in academic research the researcher needs to ask and find suitable answers to the following questions:

Pilot Studies or Field Tests for Case Study Research

1. To what extent are the research questions and sub-questions clear and answerable?
2. To what extent will it be possible to acquire access to the appropriate data to answer these questions?
3. Has a gatekeeper been found who can assist with access to informants?
4. How long is it likely to take to obtain the necessary data?
5. What forms will the data take?
6. To what extent will the researcher be able to cope with the volume of data produced?
7. How long will it take to transfer the data to a computer readable format?
8. How will data coding be performed?
9. Is there a clear strategy for data analysis?
10. Will the analysis of the proposed data lead to the research question being more fully understood and/or answered?

9.7. Feedback

The pilot study is performed to obtain feedback in order that the approach to the research can be refined and improved. It is therefore essential that the researcher takes heed of the feedback obtained. Having incorporated improvements, and if the researcher is comfortable that the challenges involved in the proposed research process can be managed, then the main research project can proceed.

9.8. Summary and conclusion

A question that is often asked is, *Can the data obtained during the pilot study be used as part of the main data acquired for the research?* If the pilot study results in no major change in data collection design then, in as far as the data obtained during the pilot study will be relevant, it may be used. In any event the pilot study is a part of the researcher's experience in getting to know his or her topic and thus it is an important part of the researcher's voyage of discovery.

The issue of pilot testing is particularly important in case study research because with this approach to research the design will inevitably be complex. Case studies always require multiple streams of data and it is not always clear whether adequately important or useful sources can be found. It is easy for an academic researcher to incorrectly believe that data will be readily forthcoming from informants, only to discover during the pilot study that this is not always the case. If this occurs then a rethink of the research design will be required. It is much more effective to rethink and re-evaluate the research design before months of work have been spent and the design has been found to be flawed.

It is not possible to estimate how long a pilot study could take but it is important not to rush this activity. Researchers who hastily conduct a pilot study run the risk of encountering potentially serious problems later in their research and perhaps having to face a substantial amount of reworking.

Chapter 10
Data Management for Case Studies – Make writing up easier

10.1. Introduction

Case study research is notorious for generating large quantities of data which can present a problem to researchers who are not prepared for the management of such a considerable amount of data. There could easily be several hundred pages of transcripts, dozens of reports from various parts of organisations, market reports, marketing literature and corporate brochures, a number of voice and perhaps video recordings, newspaper and magazine cuttings, photographs and correspondence.

Those who have read the previous chapters of this book are likely to be ready to commence their data collection activities. Irrespective of whether interviews, focus groups, questionnaires, photographs or some other approach has been chosen for data collection, the management of the data collected will become an important issue and has to be thought about carefully in advance.

Although seldom if ever offered as advice, one of the first activities academic researchers should undertake is the acquisition of a substantial filing cabinet and a systematic approach as to how to use it. And due to the fact that ethics committees today require assurances that all data obtained by the researcher from informants will be held securely, it is important that the filing cabinet acquired has a lock and key and that it is kept locked when not directly in use.

As mentioned above, with few exceptions academic research projects accumulate substantial amounts of data. This is especially true when case study research is involved because there are so many different types of data which may be collected from a case study site. The data will mainly be in the form of transcripts of interviews, field notes, focus group transcripts and notes, completed questionnaires, published reports, press clippings, as well as photographs and video and audio recordings. As a matter of routine these need to be out of sight and locked up when not in use. The researcher needs to be conscious of the possibility of data overload.

10.2. Filing systems for data saving and retrieval

Creating a filing system whereby the data collected by the researcher can be easily retrieved is fundamental to sound research practice. This involves the acquisition of a number of physical files in which the papers collected by the researcher can be systematically stored. Some of this data may well be eventually digitised and stored on computer, but there will always be the need to secure the original source documents. These original source documents are the ultimate backup for the researcher and they should be kept for some time, even after the end of the research. The researcher should check the university guidelines as to the period for which the data should be retained.

Data Management for Case Studies – Make writing up easier

There are numerous different types of filing systems and each researcher will need to choose what will suit him or her best. Some of these filing systems can be expensive while others are simple, inexpensive and easy to use.

10.3. Some details with which the researcher is probably familiar

Much of what can be said about data management for academics will be regarded by many to be common sense. However, it is a well known fact that common sense is not all that common and thus it is worthwhile using a few pages to point out some of the issues surrounding data management and making a few suggestions. Effective data management planning can prevent a number of problems occurring and facilitate an early completion of the research.

Data management may be defined in this context as putting in place procedures and practices which will prevent data loss from delaying or disrupting a research project. Of course it is not sufficient to define or simply specify these procedures and practices. They have to be conscientiously followed and this is quite difficult for many researchers. Despite the tedium involved, taking care of your work-in-progress is of prime importance. Even if you backup regularly to an offsite location, a data loss occurring the one time that you forget to do it can cause you a lot of grief – especially if it happens when you are close to submission and under pressure from deadlines.

In the production of a research dissertation a large amount of writing is necessary. In addition there will often be numerous graphs and diagrams as well as other figures and tables generated by software packages. Preparation of a dissertation will normally involve a word processor, a spreadsheet and possibly other presentation software, not to mention statistical packages such as SPSS or analysis tools such as nVivo. Some doctorates may have sufficient data to justify using a relational or other structured database. Microsoft has by far the leading position in the market for office software and thus the most likely systems will be Word, Excel and PowerPoint. It is beyond the scope of this book to address specific issues related to these products, but nonetheless it may be helpful to consider some of the data or file management issues associated with a large research project when using these tools.

10.4. Directory and file names

How documents are filed in the computer can make the researcher's work relatively easy or it can lead to confusion and the need for rework. Without proper file management documents can be lost or can be inappropriately deleted. Any reworking is wasteful and should be wherever possible avoided. The first issue here is that an easy to use directory and file naming system is required. The second issue is that it is essential to create and maintain a backup regime throughout the research project. Devising the correct structure on day one and following it carefully can save many hours of work later, so this deserves a bit of thought before you set it up.

10.5. Directories and subdirectories

The first step is to create and name a master directory for all the files related to the research project on the disk of your principal computer. Within this directory there will be the need for a number of subdirectories. The number and the content of these subdirectories will depend to a considerable extent on the type of research being undertaken. It will also de-

Data Management for Case Studies – Make writing up easier

pend on the data management style of the researcher. Planning the layout of these directories can be helpful to the smooth operation and this plan could be seen as a road map to the data required for the dissertation write up.

If empirical research is being conducted it will often be useful to have a subdirectory for the data collected and a separate subdirectory for the writing up. If different types of data are involved a researcher might wish to have one subdirectory for quantitative data which might be in Excel files and another subdirectory for qualitative data, which could be in Word files. Depending upon the software used to analyse the data more files of different types will be created and it may be appropriate to store these in different directories. As a general rule do not nest directories more than two or three times as deeply nested directories can cause data to be forgotten or mislaid.

Within the writing up subdirectory some researchers like to create another level of subdirectories, one for each chapter in the dissertation and one each for the appendices and the references (if the latter are not stored in something like Endnote). This can be useful as there will no doubt be a number of different versions of the text for each chapter and a large number of files can be confusing. It is important to incorporate a version indicator usually a number in the file title. It is hard to overemphasise the importance of version control. Losing track of which version is current can be frustrating and can waste a lot of valuable time. In a worst case scenario it can lead to errors in the dissertation.

On the question of version control, another problem may arise in having to deal with many old versions of the same file.

Good practice with computer housekeeping requires old versions of files to be deleted when new versions are in use and have been backed up correctly.

There is an art in file and directory naming. Try to design something simple, meaningful and flexible. Do not set up too many directories; keep it manageable.

10.6. Merging files or combining data

As the write up develops there may be the need to merge files with text and files with data such as tables, diagrams or graphs. On this account it is important to be aware of the size of the resulting file. The file size can grow surprisingly large and not all computers (or mail systems) operate well with large files although this is a diminishing problem. There are techniques for controlling the size of files and researchers should become familiar with these. Particular attention should be given to image files as the file type and resolution will affect both the appearance and the size of the resulting file. Furthermore, if images are copied into a text file and then cropped, it is a good idea to then cut them from and paste them back so as to lose the cropped data from the file.

As the files size grows the researcher should take more care with regular saving and backing up procedures as there is obviously more to lose.

10.7. Backing up data

Many researchers will lose some data at some time during the research project. The loss of data is painful and can even put the whole research project in jeopardy. Researchers who do

Data Management for Case Studies – Make writing up easier

not make adequate arrangements for backing up are effectively gambling with their project.

As a general rule you should backup every day during periods of moderate activity and several times a day during periods of intense writing and/or analysis. Many universities provide an automated backup service and you can use this if it is available, but during highly active periods it may not be frequent enough. If not, you should have your own backup to a reliable medium which should be stored offsite in a safe location. The minimum number of backups is one, but two is advisable. This is often referred to as a 'father and grandfather' system whereby you overwrite the day before yesterday's backup with today's backup. Where files are not too big, a simple and effective form of backup is to e-mail the files to a friend and ask them keep the copy in their mail system. You can also mail the backups to yourself although if your file sizes are large, you will need to pay attention to your assigned mailbox space. In this respect there are web based products available. One of them is called Drop Box https://www.dropbox.com/ and this or a similar service could be helpful.

Be careful in your choice of backup media. Traditionally most researchers use a secondary storage unit directly under their control with which to backup their data. Until relatively recently this might have been a CD or a DVD or some sort of fast tape device. Increasingly memory sticks, also known as flash drives, are now being used for this purpose. Although excellent in many respects these memory stick are vulnerable especially to loss, theft and failure. To overcome this, researchers sometimes keep a number of copies of their data and write ups and this can in turn lead to version identification problems (see below). If you have a large volume of data it may be worth buying a small portable hard disk drive. These are inexpensive and generally robust and they are not as prone to loss as the data sticks are.

Another approach to backing up is to use cloud based services supplied by a number of major companies globally. If you do this, the backing up process can be looked after by the outsourcer and your data will be stored in various parts of the world. On the face of it this arrangement appears to offer many advantages and can offer peace of mind to the researcher. However many countries have data protection legislation and this can prescribe that certain types of data cannot be transmitted across national borders of certain countries. Bear in mind too that while some cloud providers offer quite low cost options for moderate amounts of data storage, costs can rise steeply if you go over a certain threshold. It is not likely that many dissertations will do this, but check the fine print before signing any agreement with a cloud supplier.

Whatever approach is taken to backing up it is important to test its effectiveness from time to time. Computers are not infallible and data can be corrupted. The researcher should be reasonably confident that the data can be restored in a useful way.

Finally with regards to backing up, because computer technology now plays such an important role in research it is often forgotten that taking a paper copy of data and partial write ups is also a backup – perhaps the backup of the last resort. So take a paper copy of your work now and again.

There is much advice on this subject on the Internet by simply Googleing the term "backup".

10.8. Reference management software

Another strand of data which needs planning and control is the detailed recording of references used in the dissertation. Irrespective of whether or not empirical research is being undertaken the researcher will be required to demonstrate that a significant number of the published works of others has been read and understood. This reading needs to be recorded in a reference list and there are today a number of specialised products which can assist with this task of capturing full bibliographic details. End Notes is one of the best known. However the reference list may also be created in a word processor such as Word or any other database type product. There are helpful details on the web[28].

10.9. Computer Housekeeping

Ensuring that the data held on your computer is properly controlled and that old unnecessary versions are deleted is referred to computer housekeeping and should be part of the routine of the effective researcher. Some useful tips are available at:
http://www.pcsupportgroup.com/home-pc-support/home-pc-support-tips/home-computer-housekeeping-tips

---

[28] See: http://www.literaturereviewhq.com/6-tips-on-how-to-choose-reference-management-software/

10.10. Summary and conclusion

Data management is an important issue which has to be addressed on a routine basis. If (or perhaps when) data is lost neither the supervisor nor the University will have any sympathy for the researcher and a serious data loss could put the whole degree in jeopardy. Like so many other issues in research, with a little data management and contingency planning many tedious problems can be avoided.

The first step is to think through all the different types of data the research is likely to produce. There will be at least both physical documents and digitised documents.

The second step is to find filing cabinets and or drawer space in a desk for the secure filing of papers and recorded material.

The third step is to create a road map of how the different types of files will be stored on a directory by directory basis.

The fourth step is to create a standard for file naming with a strong emphasis on the fact that many files will have multiple versions.

The fifth step is to be aware of the implications of large files.

The sixth and last step is to decide your backup policy and ensure that these backup arrangements are satisfactory and in particular that they will work if and when they are needed.

Chapter 11
Ethics Approval for Case Study Research

11.1. Introduction

When it comes to ethics approval, case study research can offer some special challenges to researchers. The main reason for this is the need to obtain corporate or organisational consent to the research as well as the diversity of data a case study researcher may be offered when engaging knowledgeable informants at a case study site. There is also the fact that it can be difficult for a researcher to know exactly the individuals that the case study organisation might offer the researcher as informants when he or she is present on the organisation's premises. For this reason a case study researcher needs to take a particularly careful and thorough approach when applying for an ethics protocol.

11.2. Background to the importance of ethics approval

It is now well established that ethics is an important research issue. Researchers have to be cognisant of the different ways their research can impact a wide range of individuals if they are to be regarded as responsible researchers in today's complex society.

This has not always been the case. Research ethics first raised its head in the medical field at the end of the Second World War, but even then it was not taken all that seriously and it only became a real issue when Henry Beecher's whistle blowing paper was published in the New England Journal of Medicine in 1966. Beecher's paper drew attention to 22 cases in the USA where human subjects of research were abused in ways that are difficult to imagine doctors doing today. In the USA this triggered federal rules on human experimentation and informed consent. This concern spread and ethics finally became part of the general academic mind set or consciousness when David Rothman's (1991) now widely read book, Stranger at the Bedside, was published. Of course, science had been taking advantage of vulnerable people for decades, if not for centuries, but by the third quarter of the 20th century it was finally realised that this had to stop. All medical research institutions ranging from hospitals to universities were required to apply for ethics clearance before research on humans and animals could begin.

11.3. All research involving human participation

It took another 2 to 3 decades for it to be realised that the principle of ethics clearance should be applied to any research which involved human participants. Today ethics approval should be obtained before any research involving even minor contact with humans begins. This applies to all fields of study including Information Systems and all branches of management. Of course, where the research does not involve human participants, as in topics to do with pure computer science, such as algorithm optimisation or computer language performance or new designs for integrated circuits for example, then there is no need for ethical approval. As case study research involves empirical research it is almost inevitable that ethics approval will be required.

Ethics Approval for Case Study Research

It is sometimes comforting for researchers in business and management studies to know that they are not unique in being required to apply for ethics approval, which has become standard practice for all researchers who have even minimal contact with human participants such as that required when an informant completes a questionnaire or when any personal information about people is involved.

11.4. Procedural ethics

The ethics of research is a wide-ranging topic and it has implications for all aspects of the research process, as well as for the relationships between the researcher and the institution or the company in which the research takes place. In the university or research institution environment there are a number of issues related to the researcher and the supervisor, as well as concerns related to relationships between different researchers themselves. This chapter focuses on how a researcher obtains ethics approval for a proposed research project, which is sometimes referred to as procedural ethics. Therefore the issues mentioned here are highly specific to satisfying the group that oversees the research in the School/Department or organisation. There are many other ethics issues which are not raised by this process such as plagiarism, stealing ideas and deceit about the nature and source of data, and this may be referred to as ethics in practice. It is assumed by the ethics procedural process that these issues will not be transgressed by the researcher.

The issues addressed under the general heading of procedural ethics are mostly concerned with how the data will be acquired and from whom, as well as how the data will be managed and disposed of when the research is completed. The main focus of this concern is to ensure that the relationship between the researcher and informants is clear and that the researcher does not put any informant in jeopardy in any way. To this end the researcher is required to state the types of data which will be acquired and in some cases the name and position of the person from whom the data will be acquired.

Because case study research calls for multiple sources of data and because the relationship between the researcher and a number of informants may extend over a material period of time, the acquisition of an ethics protocol is a central part of case study research.

11.5. Ethics in practice

There is no easy way that ethics in practice can be effectively monitored by the institution. Many institutions are asking research degree candidates to submit their dissertations on both paper and electronically. The electronic copy of the work is then sometimes passed through a plagiarism detecting software system. Where this is done consistently it can reduce the incidence of plagiarism. But of course there is no certainty that plagiarised material will not slip though this net.

There is also the question of researchers misreporting the number of interviews which were conducted and/or the number of case studies included in the research. Some universities have begun to ask researchers to submit a report on the time spent collecting data and then performing an analysis to ensure that the data reported corresponds to the time the researcher claims to have spent on the research. This of course requires people resources from the university and at present this can be a problem. But it can be expected that there will be more activity of this nature in the future. The increasing numbers of research students and the increased cost of obtaining the degree are likely to result in some researchers not being quite as honest as they could be. These types of pressure can produce behaviour which is less than professional.

11.6. The proposed two outcomes of ethics approval

It is worth keeping in mind that the institution uses the research approval process to achieve two outcomes. The first of these is to focus the attention of the research in such a way that the researcher will not be open to any criticism on the grounds of misconduct. The second reason is that researchers, or the organisation in which the researchers work (universities or research institution), wish to protect themselves against any actions which could arise from an accusation of research misconduct.

11.7. Documents need completion

An application form for ethics approval normally has to be completed. The form will contain a number of questions related to the topic of the research, the research question, what type of participants are required, how data will be collected and so on. This means that the researcher needs to have given quite a lot of thought as to how the research will be conducted and to what extent there is any potential for ethics problems to arise. There are very few if any research projects where there is no potential for ethics problems.

11.8. Corporate research approval

Many research projects in academe involve individuals offering data in their own right. This data is often about their preferences or opinions or their experiences. In such situations the decision to supply this data is entirely up to the individual and thus only their personal informed consent is required.

However, when it comes to case study research it is often the opinions and experiences of corporate actors in their professional roles which are required. In this case the organisation needs to provide authorisation before the research may commence. Of course organisational consent does not remove the responsibility for personal informed consent, but is rather an additional layer of necessary consent.

In case study research environments Ethics Committees require details of each of the corporate participants who will take part in the research. They will also need a schedule of the individuals who will be involved in supplying the researcher with data. Some Ethics Committees require this list to contain the actual names of the people concerned. However, sometimes the role or the function of the individuals will be enough.

A list of the different types of data which will be sought should also be provided.

11.9. Ethics approval for case study research

As case study research will require approval at two levels the following is required. Firstly there will be a description of the organisation that the researcher will approach as well as details of the different areas in this organisation from which data will be sought. There will also be a list of the functions of the individuals who will be invited to contribute data. Universities may require an ethics clearance at this level before proceeding to obtain a full and detailed ethics protocol.

Ethics Approval for Case Study Research

A draft letter of corporate or organisation consent which will be signed by a senior member of the organisation who is authorised to allow the research to be conducted needs to be supplied to the Ethics Committee. An example of such a letter is provided in Figure 1.

In addition to the above the researcher will need to supply copies of the traditional ethics approval documentation.

11.10. Research Participant's Information Document

It is a normal requirement that the ethics application approval form should be accompanied by at least two separate documents. The first of these documents is what is generally referred to as a Research Participant's Information Document. This document should address all the major issues of which the informant should be made aware and in so doing it should explain the major features of the research project to the prospective informant. The form of this document, which can be seen in Figure 2, is that of a series of questions that may occur to the informant together with the answers that the researcher considers to be appropriate. It is realised that all universities or research institutions will have their own approach to the issues outlined in Figure 2, but nonetheless it is a useful indicator of the type of matters the researcher needs to think about.

11.11. A Letter of Informed Consent

The second document is a draft Letter of Informed Consent to be signed by each individual. See Figure 3 for an example of such a document. This letter confirms that the informant has received and has read and understood the Research Participant's Information Document and that any questions which he or she may have had have been answered to their satisfaction. An important aspect of this letter is the recognition that the informant is aware that he or she is not obliged to complete or answer every question asked of them, and that he or she may withdraw from the research at any time.

Where interviews are involved the researcher should have a signed Letter of Informed Consent for each informant. Where questionnaires are involved then the questionnaire itself should have a section which requires the informant to indicate that he or she has given informed consent. If the questionnaire is completely anonymous then the submission of the questionnaire to the researcher is normally accepted as an indication of Informed Consent.

11.12. Measuring instruments

In addition to these documents it will be necessary to supply a copy of any measuring instruments which will be used in the research to the ethics approval group. Research studies based on a questionnaire will need to have approval of the specific questionnaire and research based on interviews will have to show the interview schedule to the Ethics Committee. Where case studies or action research is involved a detailed plan of who exactly will supply data and how the data will be used is normally necessary.

If the research involves vulnerable people such as children, patients, the elderly, to mention only a few such groups, then additional approval may be needed for the research. If the researcher wishes to enter a school for example, additional police clearance will normally also be required.

11.13. Getting going

The body that awards ethical approval will consider these documents and decide whether the proposed research will comply with the ethics standard of the research institution or the University. It is not common for an ethics application to be approved immediately. Often the researcher or researchers will have to answer questions which are raised as a result of the consideration of their documents. Sometimes additional clarification is called for and the researcher may have to attend and provide explanations to the committee.

Sometimes the ethics approval committee will give a provisional consent to the research. In such a case the researcher will be told that he or she may commence the research project, but that he or she will be asked to report back after a month or two. For example, permission may be granted to research at one of several sites which are being requested and the researcher will be asked to consult the committee again before the second site commences. Having conducted the research at the first site the researcher may be able to give a much more comprehensive account of what is really involved in the research.

11.14. Summary and conclusion

As may be seen from the above, ethics approval is a serious matter and some research institutions and universities are very strict about to whom they will give approval. Some Universities state that if a project is started without approval it has to be discontinued and the people involved must not pursue that topic. Here no retrospective approval will be entertained. This is obviously a very extreme attitude to take.

Thus the best advice that can be given to researchers is to take ethics approval seriously and to apply for it as early as possible. It is important to expect that approval will not be given immediately and that the application may have to be resubmitted.

For a full account of this subject see:
http://academic-bookshop.com/ourshop/prod_1439563-Ethics-Protocols-and-Research-Ethics-Committees.html

Figure 1 Corporate or Organisation Informed Consent Letter

I, representing the <organisation> confirm that the <organisation> has agreed to take part in the research project being conducted by <researcher> as part of the requirements for his/her research degree <degree title> at <university>. I confirm that I am authorised to give this consent.

I have been informed about the purpose of this research and the types of data required and <organisation> is satisfied that the research will be conducted satisfactorily by <researcher>.

Any questions which I have asked have been answered to my satisfaction.

I understand that the information which <organisation> will supply is confidential and that it will be anonymised and will only be used in the findings of the research. I agree that the data may be used both in a <masters/doctoral> dissertation and also in papers arising from this research, which may be published in peer reviewed journals.

I understand that <organisation> do not have to answer all the questions which may be put to it. The information which is provided will be held securely until the research has been completed (published) after which it will be destroyed.

The information provided will not be used for any other purpose.

Ethics Approval for Case Study Research

> I understand that <organisation> is entitled to ask for a de-briefing session or a copy of the research at the end of the project.
>
> I have been informed that <organisation> may withdraw from this study at any time and that any information which supplied will not be used and any records held relating to its contribution will be destroyed. I do realise that this is only possible before the data has been anonymised.
>
> I understand that individuals who are approached for data will be asked to provide a document provided by the researcher indicating informed consent to participate in this research and that they may decline, without stating any reason, to participate in this research project.
>
> For and on behalf of the Organisation.
>
>
>
> Date
>
> Researcher
>
> Date

Figure 2 Research Participants' Information Document

The purpose of this Document is to explain to potential research participants the nature of the proposed research and the role that he or she is being invited to play in that research. ***The following is a fictitious example.***

	Issue	Detail
1	Name of researcher & contact details	Jamie Cassidy e-mail and telephone number
2	Affiliation of re-	School of Business Studies (Part-time) and

Case Study Research

	Issue	Detail
	searcher	The Hi-Tech Corporation
3	Title of Research Project	Cloud Computing, What is in it for us?
4	Purpose of the Study	The purpose of this research is to find out if Cloud Computing can advance the following: In Business; to sell and supply products or services into the cloud and the considerations of business strategies and processes to do this In IT; acquiring, and using IT/IS services provided by the cloud to do business – will it be good for ICT strategy, the IT Organisation, and the company as a whole? In the Hi-Tech sector we operate in; can the sector overall benefit by supplying to, and/or using the services of the cloud?
5	Description of the Study	The research will take the form of Interviews (external participants) and Interviews & Questionnaires (internal participants).
6	Duration of the Study	30 months
7	What is involved?	Send participants an overview of the areas they will be asked questions on when partaking in the interview.
8	Why you have been asked to participate?	You have been asked to partake in this study due to your experience in conventional and Cloud Computing, and have worked in an environment where they have been used.
9	What will happen to the information that will be given for the study?	The information will be held in a confidential manner while the work is being collated. Following the successful completion of the project all material collected as a result of the interviews and questionnaires will be destroyed. Data will be traceable back to you until it is anonymised.
10	Can you review the data after it has been written up by the researcher?	The data may be reviewed at any time before it is anonymised after which it will not be easy to identify which data came from whom.
11	What will be done with the results of the study?	The results of the interviews and questionnaires will be reported in the findings section of the research dissertation. This will be done in a completely anonymous manner.

Ethics Approval for Case Study Research

	Issue	Detail
12	What are the possible disadvantages?	I foresee no negative consequences for participants in this research.
13	In what way will the study be beneficial and to whom?	This study will provide a useful basis for companies to understand Cloud Computing (private, hybrid, or public) as an option for business initiatives, IT Service delivery (part or whole), and any associated application in the ICT sector. The study will investigate, conclude and recommend what is required (or not) to achieve this.
14	Who has reviewed this Study to ensure that it complies with all the requirements and ethical standards of the university?	The Ethics Committee of The School of Business has approved this research proposal and granted permission for this research.
15	Can permission be withdrawn having previously being granted?	Yes, all contributors shall retain the right to have their contributions to the research withdrawn at any time prior to data being anonymised.
16	Can you refuse to answer any question?	Yes, A contributor has the right to refuse to answer any question on either the questionnaire or as part of the interview and may terminate an interview at any time.

Figure 3: An example of an individual's Letter of Informed Consent

I, <informant>, agree voluntarily to take part in the research project being conducted by <XXXX> as part of the requirements for his <research degree> at <XYZ>. I have read the Research Participants' Information Document and I understand the contents thereof. Any questions which I have asked have been answered to my satisfaction.

I understand that the information I will supply is confidential and that it will be anonymised and will only be used in the findings of the research. I agree that the data may be used both in a masters/doctoral dissertation and also in papers arising from this research, which may be published in peer reviewed journals.

Case Study Research

I understand that I do not have to answer all the questions which may be put to me. The information I provide will be held securely until the research has been completed (published) after which it will be destroyed.

The information I provide will not be used for any other purpose.

I understand that I am entitled to ask for a de-briefing session or a copy of the research at the end of the project.

I have been informed that I may withdraw from this study at any time and that any information I have supplied will not be used and any records held relating to my contribution will be destroyed. I do realise that this is only possible before my data has been anonymised.

Informant

Date

Researcher

Date

Chapter 12
Evaluating your case study research

12.1. Ways of evaluating a case study

The purpose of case study research as described in this book is to facilitate the answering of a research question as part of a research degree. This research degree could be a masters or a doctorate. Also many of the issues mentioned in this book would apply to the use of case study research for the production of a peer reviewed paper which could be published in an academic journal.

There are two parts to the evaluation of case study, research and these are:-

1. Has the case study taken the form and structure described earlier in this book?
2. Has the case study facilitated the answering of the research question?

There are possibly other concerns related to evaluating a case study, such as the style of the written case study, but these are relatively less important.

12.2. Case study appropriate form and structure?

In Chapter 1 nine different dimensions of a case study were described. In evaluating a case study it is useful to use these issues as a checklist to ensure that the main factors are addressed. Here the principal concerns are whether the case study is:-

1. Used to answer complex or challenging research questions;
2. An empirical approach to answering the research question;
3. Involving many variables not all of which may be obvious;
4. Using qualitative, quantitative or mixed methods all of which can be used in either the positivist or interpretivist mode;
5. Presented as a narrative as a way of facilitating the answering of the question;
6. A clear-cut focus on a unit of analysis;
7. Recognising the context in which the research question is put and the answer is sought;
8. Not extended for a long period of time i.e. does not compete with historiography;
9. Enriched by multiple sources of data or evidence in order to offer a degree of triangulation.

12.3. Evaluating the form and structure of a case study

These nine issues could be evaluated by the researcher with or without the assistance of colleagues. However since most of these do not lend themselves to counting, which is the basis of measurement, it would not be correct to describe this type of activity as measuring a case study. It is useful to create a matrix with these issues listed as shown in Figure 8.1.

Evaluating your case study research

Issue No.	Dimension of the case study	Assessment score (1 to 9)	Remarks
1	Used to answer complex or challenging research questions		
2	An empirical approach to answering the research question		
3	Involving many variables not all of which may be obvious		
4	Qualitative, quantitative or mixed methods can be used in either the positivist or interpretivist mode;		
5	Presented as a narrative as a way of facilitating the answering of the question;		
6	A clear-cut focus on a unit of analysis;		
7	Recognised the context in which the research question is put and the answer is sought;		
8	Not extended for a long period of time i.e. does not compete with historiography;		
9	Enriched by multiple sources of data or evidence in order to offer a degree of triangulation.		

Figure 8.1: Assessment matrix for case study research

Each of the nine dimensions listed in Figure 11 requires a subjective evaluation and this needs to be followed by allocating a number in the range of 1 to 9. A score of 1 suggests little support for the presence and use of the issue and number 9 suggests strong support. Issues No. 1 to 3 would be expected to have very strong support in the order of 8 or 9. If for any reason they don't then it is possible that proper use is not being made of case study research[29].

Issue No 4 could be scored acceptably with a somewhat lower number. The reason for this is that a case study could be entirely qualitative focusing on deduction. However it is always expected that a case study would have multiple sources of data so this issue is still of some import. A score of 6 to 9 would be expected.

The last part of the matrix which contains issues 5 to 9 would be expected to have scores 8 to 9.

The column header Remarks is useful to the researcher where he or she can make comments about issues related to clarifying the importance of each line in the matrix.

Readers will recognise that this matrix is a simple checklist and that the scoring is much less important than the fact that the checklist encourages the researcher to think about each of these issues.

It is tempting to suggest that the scores in the assessment column could be summed but the resulting statistic would have only a limited use. It could be used to compare one case study to another which has been evaluated by the same evaluator.

---

[29] The scores used here are intended to be subjective and they are simply indicators for the use of the researcher him or herself. The range of 1 to 9 could be replaced by a high/medium/low assessment column.

Evaluating your case study research

12.4. Has the case study facilitated the answering of the research question?

Deciding whether the case study has contributed to answering this question is more challenging than the first question covered above.

Through the case study the researcher will have collected a large amount of data and this data will have been prepared, normally in the form of a transcript, analysed and some conclusions will have been drawn. If the conclusions drawn are regarded as an adequate answer to the research question by the researcher, his or her supervisor or supervisors and examiners, as well as other interested parties, then the case study has delivered what was intended of it. If the case study has been used for hypotheses testing then this is a relatively straightforward procedure which can be conducted and conclusions easily drawn. If any of the hypotheses have been rejected then the theory from which they were drawn will need to be modified accordingly.

Case studies are often used for theory development and in these cases the answering of the research question does not lead to a simple reject or not reject situation. In this case the answering of the research question is a step in the formulation of a theoretical conjecture but considerably more work has also to be done.

In academic research answering the research question is a necessary condition for the success of the degree but it alone is not sufficient. To comply with the criteria of necessary and sufficient, the researcher has to use the answer to the research question to add something of value to the body of theoretical knowledge and also to show evidence that the researcher should be regarded as an accomplished scholar.

At doctoral research level the student has to demonstrate that he or she has added something of value to the body of theoretical knowledge. The addition to the body of knowledge may be quite modest but nonetheless it has to be identifiable. Research degree candidates are not normally expected to develop a new theory. Rather the expectation is that they may be able to modify or extend an existing theory. Marcel Proust's (1924) comment describes well what is required from an academic dissertation at doctoral level:-

The real voyage of discovery consists not in seeking new landscapes, but in having new eyes.

One matter which needs to be addressed in this regard is that the researcher needs to create a plausible if not convincing argument that what has been achieved is worthy of being regarded as an addition to the body of theoretical knowledge. This is central to the success of the research as it is the acknowledgement that some theoretical knowledge has been created which is required for a doctorate. There is no easy way of knowing to what extent this has been achieved. Most academics would assert that academic judgement is the basis of any decision in regard to this.

This area of theory creation is subjective and as a consequence researchers need to be careful about how arguments are crafted. When the researcher moves from his or her findings to postulating a theoretical conjecture it is useful if a set of different theoretical conjectures is created. Then the researcher needs to select the one which he or she feels is the most credible or convincing and the most useful. This exercise should be completed by explaining why the other theoretical conjecture is not chosen as being convincing and as useful as the one which is selected.

On the question of scholarship it needs to be shown that the researcher has been able to bring to the case study research knowledge of the field of study as well as an understanding of the research methods available. Furthermore the researcher needs to be able to justify the use of the particular case study approach and writing up procedure they chose to use.

The conclusion of the section on the evaluation of the case study research should weigh up the plausibility of the research findings and their contribution to the body of theoretical knowledge. It would be pleasant to think that the researcher could be objective in this task but it is realised that objectivity is difficult. Therefore the researcher is asked to be reflective and to be reasonably open about his or her reflections to themselves, their supervisor and in their written work. It is believed that reflections can improve the quality of the research.

12.5. Summary and conclusion

It is normally the case that those who are skilled at academic writing will see their work more readily accepted than those who write with difficulty. As has been mentioned before, academic writing is not the same as business or personal writing and research students and early career scholars often have difficulty in expressing themselves in a satisfactory way. However with perseverance and with some help from an accomplished practitioner the skills of academic writing may be acquired. It is important to understand that this will take time and not everyone will be capable of becoming a highly regarded author. But it is important to become a competent writer. Not being an adequate academic writer is a major impediment to an academic degree.

Reference list

Agresti A and C Franklin, (2007), Statistics, p5, Pearson, New Jersey,

Bailey M, Dittrich D, Kenneally E and Maughan D, 2011, The Menlo Report: Ethical Principles Guiding. Information and Communication Technology Research, IEEE Security & Privacy, vol. 10, no. 2, September 15, pp. 71–75.

Bannister F, (2005), Through a Glass Darkly: Fact and Filtration in the Interpretation of Evidence, Electronic Journal of Business Research Methods, 3, (1), p11 – 24.

Brown B, 6 October (2011), Stories are data with a soul, TEDxTalk, http://www.youtube.com/watch?v=X4Qm9cGRub0&feature=endscreen&NR=1

Bryman A and E Bell,(2003), Business Research Methods, OUP, Oxford

Comstock G, 2012, Research Ethics : A Philosophical Guide to the Responsible Conduct of Research, Cambridge University press, Cambridge.

Gillham B, (2000) Case Study Research Methods, Continuum, London

Grbich C, (2007), Qualitative data analysis, Sage, London

Guillemin M and Gillam L, 2004 Ethics, Reflexivity, and "Ethically Important Moments" in Research Qualitative Inquiry 10: 261-280.

Gummesson E, (2000). Qualitative Methods in Management Research 2nd edition, Sage Publications, London

Hamel J with S Dufour and D Fortin, (1993), Case Study Methods, Sage, London

Heath J, (2002), Case Studies, 2nd The European Case Clearing House, Cranfield, UK

Kennedy M, (1979), Evaluation Quarterly, Vol 3, No 4, November, p 661-678

Kuhn TS, (1962), The Structure of Scientific Revolutions, University of Chicago Press, Chicago

Kuhn, T S, (1970), The Structure of Scientific Revolutions, 2nd ed., University of Chicago Press, Chicago, Il, pp. 192-93.

Leenders M, L Mauffette-Leenders and J Erskine, (2001), Writing Cases 4th Ed, Ivey, London, Ontario

Murray R and S Moore, (2006), The Handbook of Academic Writing, McGraw Hill-Open University Press, Maidenhead

Paulos J, (1998), Once Upon A Number, p14, The Penguin Press, London

Proust M, La Prisonnière Nouvelle Revue Française, Paris, 1924, Tome VI-2, 75, A la recherche du temps perdu

Punch K, (2005) Social Research Quantitative and Qualitative Approaches, Sage, London

Ray M, (1993), Introduction: What is the new paradigm in business? In The New Paradigm in Business, p5, GP Putnam's Sons, New York, 1993

Remenyi D and F Bannister (2011), Writing up your Research, Academic Publishing, Reading, UK

Remenyi D, B Williams, A Money, E Swartz, (1998) Doing Research in Business and Management, Sage, London

Remenyi D, N Swan and B van den Assem, (2011), Ethics Protocols and Ethics Committees, Academic Publishing, Reading, UK

Rothman David J, 1991, Strangers at the Bedside - A History of How Law and Bioethics Transformed Medical Decision Making, Basic Books, Pereus Book Group, NY.

Scholz R and O Tietje, (2002), Embedded Case Study Methods, Sage, Thousand Oaks

Shattuck R, (1996), Forbidden Knowledge, p9, St Martin's Press, New York

Silverman D, (1995), Interpreting Qualitative Data: Methods for Analysing Talk, Text, and Data, Sage Publications, London

Stake R, (1995), The art of Case Study Research, Sage, Thousand Oaks

Reference list

Symon G and C Cassell, 2012, Qualitative Organisational Research, p9, Sage, London

van der Blonk H, 2002, Writing Case Studies In Information Systems Research, European Conference on Research Methods, Reading University, UK

Yin R, (1989-2008), Case study research: Design and Methods, Sage, Thousand Oaks

Useful URLs

http://www.ssdd.bcu.ac.uk/learner/writingguides/1.07.htm
http://www.meaning.ca/archives/archive/art_how_to_write_P_Wong.htm
http://pgstudy.nottingham.ac.uk/apply-postgraduate-course/writing-research-proposal.aspx
http://www.gslis.utexas.edu/~ssoy/usesusers/l391d1b.htm
http://www.experiment-resources.com/case-study-research-design.html ....
http://www.experiment-resources.com/case-study-research-design.html#ixzz1qvIiBrJO
http://www.nova.edu/ssss/QR/QR3-2/tellis1.html
http://eprints.utm.my/8221/1/ZZainal2007-Case_study_as_a_Research.pdf
http://www.niehs.nih.gov/research/resources/bioethics/whatis/
http://www.socialresearchmethods.net/kb/ethics.php
http://www.who.int/ethics/research/en/
http://www.ethicsguidebook.ac.uk/
http://www.ethicsweb.eu/ere
http://www.onlineethics.org
http://www.esrc.ac.uk/about-esrc/information/research-ethics.aspx
http://the-sra.org.uk/sra_resources/research-ethics/ethics-guidelines
http://www.reading.ac.uk/internal/res/ResearchEthics/reas-REethicshomepage.aspx
http://www.ncbi.nlm.nih.gov/pmc/articles/PMC1118625/

Appendix A
A Short Note on Hypothesis Testing with Case Studies

There are researchers who would argue that hypothesis testing should not be associated with case study research. They would claim that case study research should be seen as an inductive approach more suitable for theory creation than for theory testing. Case study research is essentially qualitative and the type of data which predominates in this research paradigm will not lead itself to hypothesis testing, it is argued.

There is more than a little sense in this argument but it remains the case that some researchers claim that case studies may be used for hypothesis testing and although this may not be an ideal vehicle for this deductive type of research nonetheless an argument may be made for its use.

There are a number of misconceptions about the nature of hypothesis testing and this note addresses these specially. The main issues are:

1. What is a hypothesis test?
2. What type of claim may be tested?
3. What is the nature of the test?
4. What is the difference between the Null and the Alternative hypotheses?
5. What data is required?

What is a hypothesis test?

Hypothesis testing is a formal procedure in research whereby a researcher attempts to reject a proposition or a claim which is referred to as a hypothesis. The hypothesis or proposition should ideally have been deduced from an established theory, although in practice hypothesis testing is used more generally and thus can be based on a researcher's judgment as well as a theory.

What type of claim may be tested?

For a hypothesis to be testable it is necessary that it is bounded in the sense that it is possible to acquire data which could be used in a suitably constructed test. Thus the proposition that men are paid more than women would not be regarded as a researchable hypothesis, at least not as a hypothesis which may be formally tested. However, a hypothesis that male and female teachers in the UK are not paid equally throughout their careers could be formulated as a suitable hypothesis for which data could be found and thus could be tested.

What is the nature of the test?

A hypothesis is tested by the researcher collecting an appropriate set of data relating to the claim made by the hypothesis. In formal statistical hypothesis testing the data should have been collected by means of random sampling and the population from which the data has been collected should have been normally distributed.

A Short Note on Hypothesis Testing with Case Studies

What is the difference between the Null and the Alternative hypotheses?

The normal structure of a hypothesis test is that there needs to be two statements or claims. The first statement is referred to as the Null Hypothesis and is normally designated by the symbol H_0. The second statement is referred to as the Alternative Hypothesis and is normally designated as H_a. These two hypotheses are coupled because if the researcher is able to reject the Null Hypothesis then the Alternate Hypothesis is 'accepted' pro tem. It is important to note that a hypothesis is never proved or believed to be 'true'. A hypothesis can only be rejected or not rejected. If the Null Hypothesis is not rejected then it is 'accepted' pro tem.

What data is required?

Hypothesis testing is normally associated with quantitative research. To test a hypothesis data is collected from a sample of a population of interest. By knowing the size of the sample and by knowing the measure of central tendency and the appropriate measure of dispersion in the data a statistical test may be constructed. With this approach the researcher can not only reject a hypothesis, but can also determine the level of confidence associated with the outcome of the test. Standard statistical techniques are available for this purpose.

The term hypothesis testing is sometimes used loosely in qualitative research although the term proposition testing is probably considered more appropriate. When this type of hypothesis testing is done then the researcher attempts to find non-quantitative evidence which will allow the Null Hypothesis or proposition to be rejected. But this evidence and thus test will not be statistical. The researcher will normally through interviews, focus groups, observation or archived documents seek data which will support the contention that the proposition is rejected. No statistical level of confidence can be associated with this type of test. Nonetheless it has become acceptable for qualitative researchers to seek out appropriate data and to conduct tests on it as described here. It is however important that when hypothesis testing is used in this manner that the researcher makes it known that this informal approach has been used. It is not possible to say how many data items a researcher would have to collect in order for him or her to claim that a hypothesis could be rejected in this fashion. This is a subjective matter. However a reasonable number of data items would be needed in order to give any such statement even minimal credibility.

In general, this technique may be seen as an operationalisation of Popper's Falsification concept in that it focuses on the rejection of the Null Hypothesis.

Index

After data collection is complete, ii, 71, 75
Alternative hypothesis, 201
Assessment matrix, 191
Audit trail, 88, 153, 154
Backup, 76
Bottom up, 76
Boundaries, 3
Case study site, 117
Coding, 76, 77
Combining data, 168
Computer housekeeping, 168, 171
Concept flow charts, 123
Concept profile, 127
Contemporary phenomenon, 3
Content analysis, 110
Control, 10
Correspondence analysis, 75, 110, 111, 120, 121
Credibility, 21
Cross case analysis, iii, 102
Data access, iii, 93
Data acquisition plan, 52, 53
Data caches, 48, 51
Data management, 166, 173
Data overload, iv, 118
Data protection legislation, 170
Data stick, 170
Deadlines, 165
Demonstrating the relationships, iii, 109, 112
Digitised, 164, 172
Directories, v, 167
Do, dare, dedi, datum, ii, 36
Elicited data, 43
Embedded case studies, 26
Empirical enquiry, 3
Establishing the data required, 71, 76
Ethics, vi, 73, 84, 175, 178, 180, 181, 182, 184, 187, 197, 198
Ethics approval, vi, 180
Ethics committee, 70, 75, 151, 164, 178, 179, 185
Ethics in practice, vi, 178
Ethics protocol, 84

Ethnographic, 48, 49
Evaluation, 197
Evidence, 36, 37, 197
Executing the research design, 66
Fact, 197
Falsification, 204
Father and grandfather, 169
Field note, 48, 49, 97, 100
Field test, 43, 100, 130, 157
File Management, 166
Filing, v, 165
Finding the starting point, ii, 71
First contact, iii, 91
Flexibility, 77
Focus group, 49, 96
Gatekeeper, 73
Generalisability, 132
Human participant, 174, 175
Hypothesis testing, iii, 6, 113, 202, 203
Hypothesis, iii, vi, 6, 113, 201, 202, 203
Informants, 73
Informed consent, 174, 178, 179, 180, 182, 183, 185
Large files, 168

Literature review, 61, 63, 71
Log, iii, 88
Magazine cuttings, 163
Measuring instrument, vi, 182
Merging files, v, 169
Multiple sources of evidence, 4
Narrative, 120
Natural data, 43
Nature of hypothesis, 199
Necessary data, 51, 90, 161
Null Hypothesis, 203, 204
Observation data, 97
Observation, i, 1, 48, 97, 98
Opportunistic, iii, 88
Participant observer, 98
Photographs, 49, 51, 90, 163, 164
Physical artefacts, 48, 50
Pilot study, 100, 101, 157, 158, 159, 160, 161
Planning, 57, 84, 168
Possible findings, 61, 67
Possible limitations, 62, 67
Post hoc ergo propter hoc, 11
Press clippings, 164
Procedural ethics, vi, 177
Proposition, 200, 201, 202
Qualitative research, i, 5

Index

Questionnaire, 93
Real life context, 3
Reference management, v, 172
Research design, 61, 64
Research ethics, 175
Research participant information document, 179
Research question, 19, 53, 72
Rework, 137, 164, 166
Scholarship, 193
Scientific logic, vii, 9
Sensitising a researcher, 51
Servqual, 41
Skill, 104
Software, 58, 59, 66, 165, 166, 167, 171, 176
Spoken evidence, 96
Statistics, 45, 197
Subdirectories, 166

Surveys, 13
Teaching – learning case study, i, 15
Theft, 169
Themes, 78
Theoretical conjecture, iii, 113
Top down, 76
Transcript, 76
Transferability, 21
Triangulation, iii, 98, 99
Unit of analysis, ii, 37
University guidelines, 164
Version control, 76
Versions, 79, 155, 167, 168, 171, 172
Work-in-progress, 165
Writing up, iv, 118, 136, 137, 198
Written evidence, 96